T0324749

The Scythrididae (Lepidoptera)
of Northern Europe

SWEDEN

Sk.	Skåne	Vrm.	Värmland
Bl.	Blekinge	Dlr.	Dalarna
Hall.	Halland	Gstr.	Gästrikland
Sm.	Småland	Hls.	Hälsingland
Öl.	Öland	Med.	Medelpad
Gtl.	Gotland	Hrj.	Härjedalen
G. Sand.	Gotska Sandön	Jmt.	Jämtland
Ög.	Östergötland	Äng.	Ångermanland
Vg.	Västergötland	Vb.	Västerbotten
Boh.	Bohuslän	Nb.	Norrbotten
Dlsl.	Dalsland	Äs. Lpm.	Äsele Lappmark
Nrk.	Närke	Ly. Lpm.	Lycksele Lappmark
Sdm.	Södermanland	P. Lpm.	Pite Lappmark
Upl.	Uppland	Lu. Lpm.	Lule Lappmark
Vstm.	Västmanland	T. Lpm.	Torne Lappmark

NORWAY

Ø	Østfold	HO	Hordaland
AK	Akershus	SF	Sogn og Fjordane
HE	Hedmark	MR	Møre og Romsdal
O	Opland	ST	Sør-Trøndelag
B	Buskerud	NT	Nord-Trøndelag
VE	Vestfold	Ns	southern Nordland
TE	Telemark	Nn	northern Nordland
AA	Aust-Agder	TR	Troms
VA	Vest-Agder	F	Finnmark
R	Rogaland		

n northern   s southern   ø eastern   v western   y outer   i inner

FINLAND

Al	Alandia	Kb	Karelia borealis
Ab	Regio aboensis	Om	Ostrobottnia media
N	Nylandia	Ok	Ostrobottnia kajanensis
Ka	Karelia australis	ObS	Ostrobottnia borealis, S part
St	Satakunta	ObN	Ostrobottnia borealis, N part
Ta	Tavastia australis	Ks	Kuusamo
Sa	Savonia australis	LkW	Lapponia kemensis, W part
Oa	Ostrobottnia australis	LkE	Lapponia kemensis, E part
Tb	Tavastia borealis	Li	Lapponia inarensis
Sb	Savonia borealis	Le	Lapponia enontekiensis

USSR

Vib   Regio Viburgensis   Kr   Karelia rossica   Lr   Lapponia rossica

FAUNA ENTOMOLOGICA SCANDINAVICA
Volume 13   1984

The Scythrididae (Lepidoptera) of Northern Europe

by

Bengt Å. Bengtsson

E. J. Brill/Scandinavian Science Press Ltd.

Leiden · Copenhagen

Fauna entomologica scandinavica
is edited by "Societas entomologica scandinavica"

Editorial board
Nils M. Andersen, Karl-Johan Hedqvist, Hans Kauri,
N. P. Kristensen, Harry Krogerus, Leif Lyneborg,
Hans Silfverberg

Managing editor
Leif Lyneborg

World list abbreviation
Fauna ent. scand.

Printed by
Vinderup Bogtrykkeri A/S
7830 Vinderup, Denmark

ISBN 90 04 07312 4
ISBN 87 87491 21 4
ISSN 0106-8377

Author's address:

Fil. mag. Bengt Å. Bengtsson
Box 71
S-380 74 Löttorp
Sweden

Contents

Abstract

This treatment covers all species of Scythrididae known from Denmark, Finland, Norway, Sweden, Great Britain, Holland and Poland, and also most of the species from Germany, Belgium, Czechoslovakia and from western Russia north of 50° latitude. All 35 species are illustrated in full colour and supplementary illustrations of some infrasubspecific forms are provided; male and female genitalia are also figured. The wing venation of several species is shown. All species are keyed and redescribed. A synopsis of the morphology, immature stages and bionomics of the family Scythrididae are provided. The zoogeography, systematics and classification are discussed.

The following species are dealt with (asterisks denote species known from Fennoscandia and/or Denmark):

* 1. *Scythris obscurella* (Scopoli, 1763)
2. *S.cuspidella* ([Denis & Schiffermüller], 1775)
* 3. *S.potentillella* (Zeller, 1847)
* 4. *S.cicadella* (Zeller, 1839)
5. *S.bifissella* (Hofmann, 1897)
* 6. *S.limbella* (Fabricius, 1775)
* 7. *S.knochella* (Fabricius, 1794)
8. *S.scopolella* (Linnaeus, 1767)
9. *S.paullella* (Herrich-Schäffer, 1855)
10. *S.clavella* (Zeller, 1855)
11. *S.seliniella* (Zeller, 1839)
12. *S.subseliniella* (Heinemann, 1877)
13. *S.sinensis* (Felder & Rogenhofer, 1875)
* 14. *S.productella* (Zeller, 1839)
* 15. *S.palustris* (Zeller, 1855)
16. *S.muelleri* (Mann, 1871)
* 17. *S.inspersella* (Hübner, 1817)
* 18. *S.noricella* (Zeller, 1843)

* 19. *S.empetrella* Karsholt & Nielsen, 1976
* 20. *S.siccella* (Zeller, 1839)
21. *S.tributella* (Zeller, 1847)
* 22. *S.picaepennis* (Haworth, 1828)
* 23. *S.disparella* (Tengström, 1848)
* 24. *S.fuscopterella* Bengtsson, 1977
25. *S. braschiella* (Hofmann, 1897)
* 26. *S.laminella* ([Denis & Schiffermüller], 1775)
27. *S.crassiuscula* (Herrich-Schäffer, 1855)
* 28. *S.ericivorella* (Ragonot, 1881)
* 29. *S.crypta* Hannemann, 1961
30. *S.restigerella* (Zeller, 1839)
31. *S.dissimilella* (Herrich-Schäffer, 1855)
* 32. *S.fuscoaenea* (Haworth, 1828)
33. *S.grandipennis* (Haworth, 1828)
34. *S.ericetella* (Heinemann, 1872)
35. *S.fallacella* (Schläger, 1847)

Introduction

The present work deals with 35 north-western European species of the lepidopterous family Scythrididae. The scope of the treatment has been limited by practical considerations to the exclusion of the large number of species of Scythrididae in central Europe.

The Scythrididae is one of the few families of 'Microlepidoptera' that has not yet been revised during the twentieth century. Owing to the similarity of external characters, many synonyms have been produced.

The only attempt to review comprehensively the current knowledge of Scythrididae was made by Zeller (1855). Since then only limited faunistic articles have been published; however, during the last few decades some revisions of collections in different museums have clarified several nomenclatural problems.

All available literature concerning the species dealt with here has been examined. Due to historical difficulties of determination, however, only certain information thought to be reliable is adopted, unless the records or observations are of such special interest that they might provoke confirmatory studies.

All known synonyms are listed in chronological order, with a reference to the original description.

The redescription of each species comprises size and appearance of the moth (male and female are described separately if they differ), a diagnosis, and a description of the male and female abdomen which in most cases offers useful specific characters. Further, the genitalia of male and female are described. The known geographical distribution is given, many times based on unpublished material. As in earlier works in this series, the detailed distribution in Fennoscandia and Denmark is shown in tabular form.

Immature stages, foodplants, phenology and habitats are described under the heading 'Biology'. Names of foodplants follow Lid (1974).

Because of difficulties in determination of, in particular, the unicolorous species, I have limited the descriptive text. Due to these difficulties the key based on external characters should be considered as a rough guide only; identification would be confirmed by the dissection of genitalia.

The identification of the different components of the genitalia often demands a careful examination of the fusion of the sclerites and this has not been accomplished for all species.

During the compilation of this volume many workers have given me much information and support. I am greatly indebted to Mr. Eberhard Jäckh, Hörmanshofen, for providing material, for valuable discussions and useful information about the distribution of several European species. I owe Mr. Ole Karsholt, Zoologisk Museum,

Copenhagen, and Mr. Ingvar Svensson, Österlöv, a great debt of gratitude for information about many Danish and Swedish species respectively, and for the loan of material; Mr. Karsholt also gave valuable comments on some nomenclature problems.

Further, I wish to thank the following for valuable information and loans or gifts of material which facilitated the treatment of many taxa: Mr. L. Aarvik, Ås; Mr. G. Baldizzone, Asti; Mr. J.-O. Björklund, Borlänge; Dr. J. Buszko, Toruń; Mr. F. Coenen, Brussels; Prof. L. A. Gozmány, Budapest; Prof. H. J. Hannemann, Museum für Naturkunde an der Humboldt-Universität zu Berlin; Dr. P. Ivinskis, Vilnius; Mr. J. Jalava, Zool. Mus., Helsinki; Dr. F. Kasy, Vienna; Dr. J. Kyrki, Zool. Inst., Oulo; Mr. G. R. Langohr, Simpelveld; Dr. J. Patočka, Res. Inst. For. Protection, Zvolen; Mr. J. Roche, Chesterfield; Mr. A. Šulcs, Riga.

I am also grateful to Mr. Bert Gustafsson, Naturhist. Riksmus., Stockholm; Dr. E. Schmidt Nielsen, previously Zool. Mus., Copenhagen (now CSIRO, Canberra), and Dr. B. J. Lempke, Inst.Tax. Zool., Amsterdam, who generously provided loans of important material.

I also thank Mr. Roland Johansson, Växjö, for valuable advice concerning the technique of aquarelle painting and providing materials.

Dr. P. Passerin d'Entrèves, Mus. Ist. Zool. Sist. Univ. de Torino, kindly clarified some nomenclaturel problems.

I have received information on the distribution of Scythrididae from Dr. M. I. Falkovitsh, Zool. Inst. Acad. of Science, Leningrad; Prof. E. Niculescu, Bucarest; Dr. J. O'Connor, Nat. Mus. of Ireland, Dublin, and Prof. D. Povolny, Česk. Akad.Věd, Brno.

Dr. G. S. Robinson, British Museum (Natural History), Dr. N. P. Kristensen, Copenhagen (editor of Lepidoptera section, Fauna ent. scand.) and Dr. L. Lyneborg, Copenhagen (managing editor, Fauna ent. scand.) corrected the language and proposed several large-scale improvements for which I owe them a great debt.

I express my sincere gratitude to all the other people who have supported me during this work.

Finally, I give my sincere thanks to the board of the foundation of Larsénska fonden and to the Committee of Längmanska kulturfonden for financial assistance.

Family Scythrididae

Scythrididae Spuler, 1910: 432.
Elachistidae Bruand, 1850: 15.
Butalidae Wocke, 1877: 436.
Scythridae Meyrick, 1928: 722.

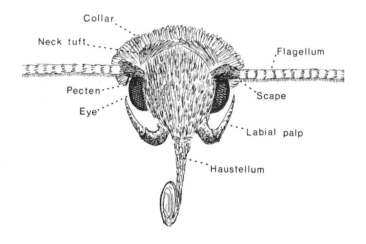

Fig. 1. Head of *Scythris.*

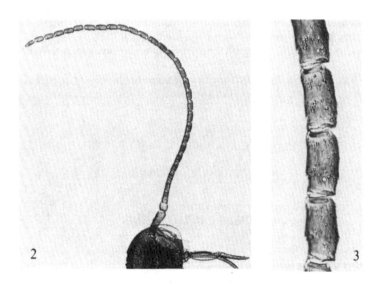

Figs. 2, 3. Antenna of *Scythris laminella* (Den. & Schiff.).

12

Morphology of adults

Head (Figs. 1-3)

Rather small; comparatively broad, flush with thorax; smooth-scaled, but sometimes with two groups of long transverse scales as an extention of neck tufts; frons vaulted; eyes rather small; with or without ocelli; chaetosemata absent.

Maxillary palpi very small, directed downwards. Labial palpi three-segmented, slender, with appressed scales, rather long (usually 2-3 times greatest diameter of eye), diverging, curved and ascending, often reaching height of vertex; terminal segment pointed, shorter than second segment. Haustellum with more or less appressed scales.

Antennae shorter than forewing (50 to 75%); smooth, sometimes weakly serrated; in male with short cilia, in female cilia considerably shorter or almost absent; without eye-cap; scape generally with pecten. Segments of flagellum unicolorous.

Thorax

Patagia well developed, with appressed regular scales.

Wings (Figs. 4-12)

Forewing elongate, broadly lanceolate; costa usually slightly bent before apex; three to five times as long as broad. Hindwing more or less sharply pointed, narrower than forewing, from half as broad to as broad as forewing. Cilia of hindwing longer than width of wing.

A brachypterous species has recently been discovered in the USA (Powell, 1976).

Forewing retinaculum on subcosta; frenulum with one bristle in male and with two or (generally) three bristles in female.

Forewing Sc reaching costa before middle, occasionally with a rudimentary basal branch; R five-branched, R_1 arising in or after middle of wing, seldom before middle (Hering, 1918), R_1-R_4 to costa, R_5 to termen (cf. Hering, loc. cit.), R_4 and R_5 stalked; M three-branched, M_3 usually coalescent with CuA_1; the first phase in this process of fusion is seen in *S.fallacella* (Schl.) (Spuler, 1892). Analis single and weak, becoming stronger at tornus.

In hindwing $Sc + R$ and Rs parallel, the former relatively long, extending beyond middle of costa, the latter in a few cases stalked with M_1 (Fig. 9); M usually with three separate branches, but often M_2 and M_3 stalked or coalescent. In species with narrow hindwing, transverse vein may be missing.

The variation of venation in Old World species is great; de Joannis (1920b) has pointed out that variation may occur within one species, occasionally in the same specimen. For example, *S.inspersella* (Hb.) normally shows all three median veins but sometimes M_3 is missing (Fig. 8).

In some species the veins are considerably thickened and strong, e.g. *S.potentillella* (Z.) (Fig. 4) while in others *(S.limbella* (F.), *inspersella* (Hb.))* the venation is comparatively weak.

Legs (Figs. 13-15)

Fore tibia with long epiphysis, about one third length of tibia, sometimes longer. Mid tibia with pair of apical spurs. Hind tibia with two pairs of spurs; medial ones beyond middle, apical spurs of unequal length, the outer shortest. Tarsal spines thinly scattered on dorsal surface, not only at articulations. Tibia dorsally and ventrally with long hairs or bristles.

Abdomen

Generally broad and heavy, in female so gross as to affect flight, the more slender male abdomen almost unicolorous. Anal tuft of varying shape, e.g. oval and curly or trifid, constituting a useful attribute for determination.

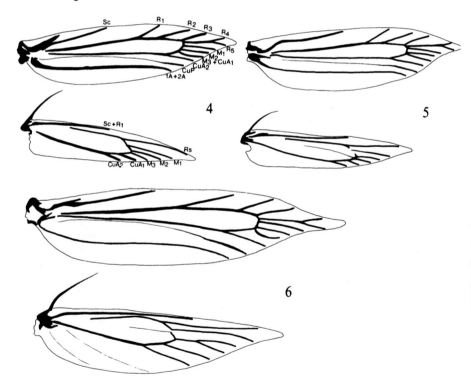

Figs. 4-6. Wing venation of – 4: *Scythris potentillella* (Z.); 5: *S.cicadella* (Z.); 6: *S.limbella* (F.).

During the nineteenth century, entomologists based their determination of *Scythris* to a great extent upon the coloration of the abdomen. The pattern of the ventral surface is frequently species-specific, particularly in females. Usually the anterior segments are dark while the penultimate segments are of a paler tone, often conspicuously contrasting white, yellow or ochreous.

The genitalia of Scythrididae are of very complex structure. Often, species with very similar external appearance have quite different genitalia. On the other hand, some species are very different in colour, markings and size but, in spite of that, show great

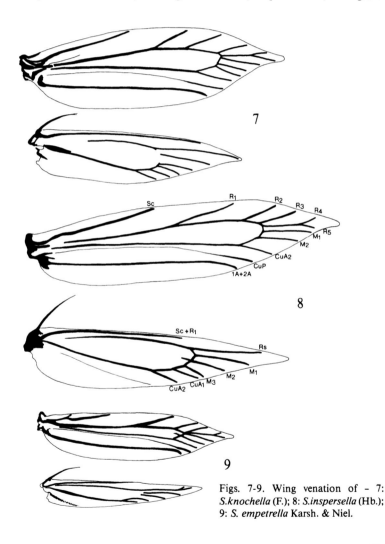

Figs. 7-9. Wing venation of – 7: *S.knochella* (F.); 8: *S.inspersella* (Hb.); 9: *S. empetrella* Karsh. & Niel.

15

affinities in genitalia. Quite often males and females are found not to fit in one and the same morphological aggregation.

Male genitalia (Figs. 16, 17)

Extremely heterogenous, in some species exhibiting asymmetry, abnormal development, fusion, reduction, loss or functional change in the different components. In many species the genitalia pose considerable difficulties in interpretation.

Tergum VIII and sternum VIII frequently offer useful characteristics for determination and form a functional part of the genitalia. Tergum VIII is usually less predominant than sternum VIII which is often more or less triangular with a proximal furcation or showing an indication of such a furcation.

Uncus normally furcate or single, but sometimes absent. Gnathos in most cases present, usually pointed and curved, sometimes labiate, laminated or ring-shaped; occasionally absent. Tegumen of normal construction or transformed. Valva of variable form, in most species well developed, often curved and tapered, in others compact,

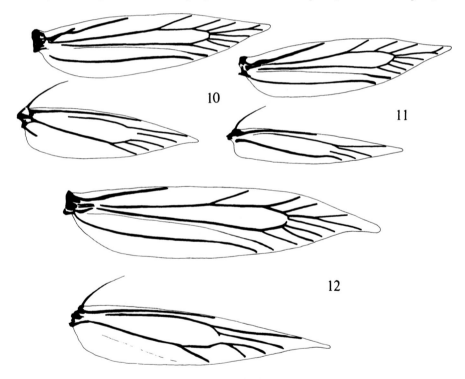

Figs. 10-12. Wing venation of - 10: *S. picaepennis* (Hw.); 11: *S.laminella* (Den. & Schiff.); 12: *S.fuscoaenea* (Hw.).

accreted or even almost lost. Aedeagus generally long, tubiform, curved and tapering but sometimes short, thick, straight or, occasionally, of diminutive size.

Zander (1905) was the first to elucidate the strange construction of tergum VIII and sternum VIII which ostensibly form a functional part of the genitalia.

In *S.noricella* (Z.) tergum VIII and sternum VIII are fused together (Fig. 16) while in *limbella* (F.) segment VIII is distinctly bisected (Fig. 17).

In the majority of *Scythris* spp. the valvae are easily identified. Sometimes, however, they are of peculiar shape. Thus *inspersella* (Hb.) has slender and strongly sclerotized valvae, fused ventrally at the base and continuously united with the lateral processes (Fig. 79). Homologous lateral processes seem to be present in *scopolella* (L.) and there, too, certain asymmetry is obvious (Fig. 70). Otherwise, asymmetry is very clearly expressed in some species with compact genitalia (e.g. *siccella* (Z.), Fig. 82) and notably in some south European *Scythris* (s.l.) belonging to the anomalous *acanthella*-group.

In *S.picaepennis* (Hw.) a ventral view of the end of the abdomen (Figs. 84, 88) conveys the impression of very simplified but intricate genitalia. In lateral view, however, most parts of genitalia and segments will be discernible.

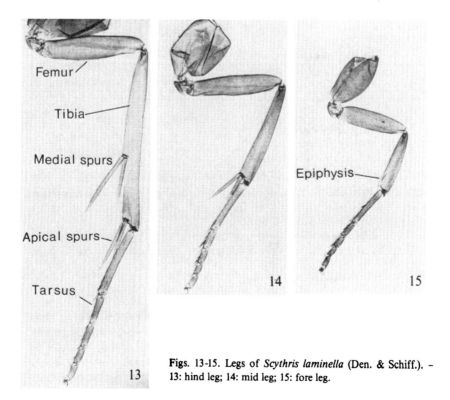

Figs. 13-15. Legs of *Scythris laminella* (Den. & Schiff.). – 13: hind leg; 14: mid leg; 15: fore leg.

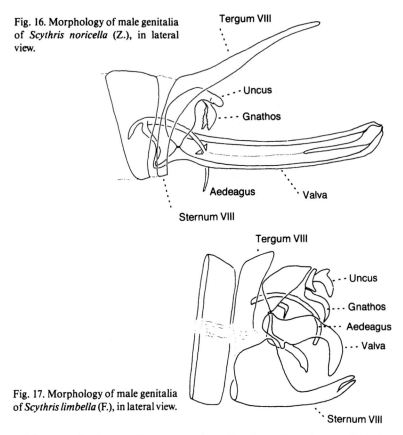

Fig. 16. Morphology of male genitalia of *Scythris noricella* (Z.), in lateral view.

Tergum VIII

Uncus

Gnathos

Aedeagus

Valva

Sternum VIII

Tergum VIII

Uncus

Gnathos

Aedeagus

Valva

Fig. 17. Morphology of male genitalia of *Scythris limbella* (F.), in lateral view.

Sternum VIII

Many species have rudimentary valvae. The *'grandipennis*-group' (Jäckh, 1977) constitutes a representative example of this condition (Figs. 99-100). The predominant tergum VIII and sternum VIII confine the very small genitalia with diminutive valvae that are only small folds with fine bristles.

In Scythrididae the aedeagus is often of differing shapes. In the main it appears in two forms, as a long, more or less bent and tapered tube or as a small, bottle-shaped organ.

Female genitalia (Fig. 18)

Papillae anales bilobed, weak and with bristles; apophyses posteriores long, about three times length of papillae anales. Tergum VIII normally of conventional type, sometimes slightly modified; apophyses anteriores in general ordinary but occasionally of aberrant form. Lamella postvaginalis smooth or granular, often of specific form, generally ringshaped or triangular. Ostium bursae in many species bordered

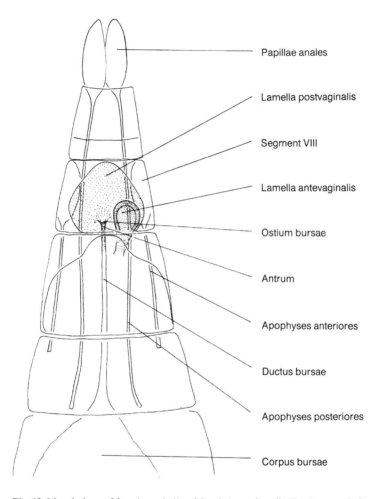

Fig. 18. Morphology of female genitalia of *Scythris productella* (Z.), in ventral view.

cephalad and ventrad by a spatular, mobile disc, the lamella antevaginalis, which frequently is indistinct. Antrum funnelled or tubular, visible in some species but usually very ill-defined. Corpus bursae and cephalad part of ductus bursae weakly membranous, in preparation only appearing when dissected and stained. No species exhibit a signum or other sclerotization of the corpus bursae.

Sternum VII more or less shield-shaped, with modified caudal margin by which the female is reputedly held during copulation (Falkovitsh, 1981). Sometimes a similar modification is present on sternum VI (Falkovitsh, loc.cit.).

Immature stages

Very little is known of the immature stages of Scythrididae; the literature gives mostly uncertain information about foodplants and pupation. Many old observations have not been verified subsequently, possibly because frequent misidentifications now make it difficult to ascertain which species was referred to.

Eggs

No detailed description of the geometry and surface structure of Scythrididae eggs has been made. Schütze (1904) reared *S.palustris* (Z.); the eggs were cream, elongate, often of irregular form, faintly glossy, and with a granulated surface. After some days the yellowish colour increased and immediately before hatching the eggs became reddish.

This description agrees well with observations made by the present author. In captivity, a female *S.laminella* (Den. & Schiff.) laid eggs on the glands of *Hieracium pilosella* L. near the leaf base. During the first few days they were pale greenish yellow but later they darkened somewhat and assumed a pale orange tone. The eggs were ellipsoid, of upright type, with a flattened base and top. The chorion showed a slightly waved network of rounded ridges, roughly consistent with the structural type illustrated in fig. 97 of Döring (1955). The micropyle was rather indistinct.

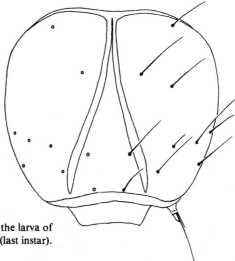

Fig. 19. Head capsula of the larva of *Scythris inspersella* (Hb.) (last instar).

Larvae (Figs. 19-21)

Most larvae of Scythrididae live concealed, often residing in sparse webs near or on the ground, or close to a leaf or the stem of the hostplant. They feed externally on leaves, buds or, occasionally, flowers (Meyrick, 1928; Schütze, 1931; Falkovitsh, 1981, etc.). When disturbed they quickly return to a safer place in the web gallery. Larvae of several species (e.g. *S.limbella* (F.) and *inspersella* (Hb.)) live gregariously and only become solitary during the last instar.

Descriptions of the larvae in literature often are only briefly outlined (e.g. Schütze, 1897; Benander, 1965 and Sattler, 1981). The larva is spindly, with a well developed prothoracic shield and anal plate. The integument is fairly smooth and the coloration exceedingly variable, sometimes with scattered pale or dark minute spots. Longitudinal lines are in most cases present.

Head capsule (Fig. 19) rounded, with straight border to occipital foramen; setae relatively short, punctures indistinct; antennae short, with basal setae; spinneret elongate; palps rather prominent with seta on the first and second segment.

Thoracic legs with straight tarsal claw and shorter, lateral process. Abdominal prolegs with crochets arranged in a partial quadriordinal circle (Fig. 20).

MacKay (1972) examined the chaetotaxy of the Nearctic *Scythris epilobiella* McD. and found it to agree well with that of the Blastobasidae. She indicated the following 'probable family characters' shared by the two taxa: 'three prespiracular setae on prothorax; L1 and L2 on meso- and meta-thorax together and L3 distant from them; L1 and L2 together on segments 1-8; D1 on segment 9 distinctly anterior to a straight line joining D2 and SD1; the SD1 seta on segment 8 hair-like, dorsal and somewhat posterior to spiracle, and apparently lacking the usual base'. And additional characters similar to those of Blastobasidae are: 'L2 on segments 1-8 very close and dorsal to L1; SD1 on segments 1-7 with base set in a conspicuous pale area ringed with dark brown; SD1 on segment 9 more delicate than nearby setae but with base apparent. The posi-

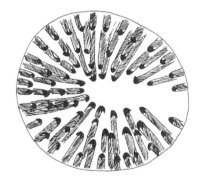

Fig. 20. Schematic arrangement of crochets on proleg in larva of *Scythris inspersella* (Hb.) (quadriordinal circle type).

tion of SD2 on segments 1-8 could not be observed. No depression or hollow, as in Blastobasidae, is present on the submentum of late-instar larvae'.

In the present work the chaetotaxy of a single Palearctic species *(S.inspersella* (Hb.)) has been studied in some detail (Fig. 21); primary setae and some secondary setae are indicated in the sketch. On prothorax XD and anterior D and SD setae arranged in vertical row as are also posterior D and SD; L setae in horizontal row; SV on common pinaculum; MV and V close to coxa. On mesothorax D in middle; SD1 and SD2 more anterior; SV on common pinaculum, one remote. Abdominal segments with large SD above spiracle, surrounded by round pinaculum and distinct, pale ring. L setae in long, oblique sequence. Proprioceptors seem to be absent. Several seta groups comprise prominent secondary setae; this is true also of *S.noricella* (the chaetotaxy of which was briefly outlined by de Lesse & Viette, 1949) as well as of the Nearctic *Scythris* studied by MacKay and the Hawaiian Scythridid figured by Zimmerman (1978).

It has been suggested that the position of abdominal segment SD just above the spiracle and the pale ring around the pinaculum are generic characters (Benander, 1965). Another common characteristic may be the placement of the SV setae on the abdominal segments. However, detailed investigation of many species is required before definite family (and generic) features can be identified.

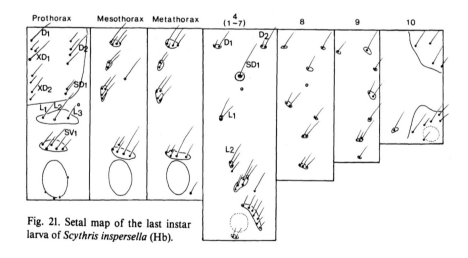

Fig. 21. Setal map of the last instar larva of *Scythris inspersella* (Hb).

Pupae (Fig. 22)

General characteristics have been outlined by Mosher (1916) and Common (1970). The pupa is elongate, somewhat flattened, broadest at the thorax, without dorsal abdominal spines. Only the last two or three segments are movable which implies non-protrusion of the pupa at ecdysis. Further family characters are suggested by Mosher

22

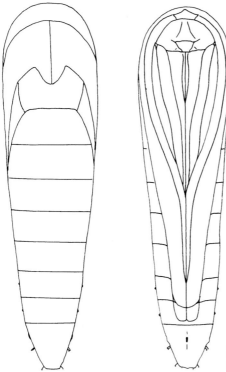

Fig. 22. Pupa of *Scythris noricella* (Z.). – to the left in dorsal view; to the right in ventral view.

(1916); femora of prothoracic legs never visible; maxillary palpi minute; tenth abdominal segment with hooked setae at caudal end.

Pupation occurs in a thin cocoon. Further observations are given for individual species.

Bionomics

Scythrididae are small, smooth-scaled moths, the northern European species usually with dark colours and few markings. They are diurnal with a few exceptions but in subtropical and tropical regions many *Scythris* species tend to be pale and nocturnal, often attracted to light. The adults are rarely seen but rest in flowers or on low, erect vegetation; they are reluctant to fly. When disturbed they mostly jump to the ground and lie motionless. Adults may be collected by sweeping vegetation but are difficult to see among debris in the net. Careful examination of the sweepings may reveal the drop-shaped moths feigning death; however, if sunshine falls on them they will try to escape by jumping and making short flights.

In Fennoscandia and Denmark most species are univoltine but a few may be bivoltine. In any case they have a very extended flight period. Hibernation occurs as an egg, larva or pupa.

Systematics and Classification

The taxonomic position of the Scythrididae has been unstable in the past, partly due to the group exhibiting a mixture of 'conventional' characteristics, and to the limited systematic treatment it has received. The family name Scythrididae (or sometimes Scythridae) was, during the nineteenth century, practically equivalent to the genus *Scythris (Butalis)* as we recognize it today. Thus the history of the family is also that of the genus.

Stainton (1854) and Zeller (1855) considered *Scythris* (or *Butalis*) to belong to the Gelechiidae. Meyrick (1895) placed *Scythris* in the Elachistidae, which placement also was adopted by Rebel (1901). Forbes (1923) assigned *Scythris* to Yponomeutidae; Barnes & McDunnough (1917) and McDunnough (1939) were of the same opinion, including the genus in the Yponomeutoidea.

Spuler (1910) introduced the family-group name Scythrididae for three subfamilies (Epermeniinae, Scythridinae and Amphisbatinae) and placed Scythrididae between Elachistidae and Yponomeutidae. This was to a large extent followed by Hering (1932).

Meyrick (1928) revised his earlier views and erected the new family Scythridae, placing it between Elachistidae (-Douglasidae) and Hyponomeutidae, in the Hyponomeutoidea.

Meyrick and Spuler paid great attention to wing venation in classification and this phenetic taxonomy was also embraced by Turner (1947). The latter author reviewed briefly previous classifications and tried, in phylogenetic terms, to make a new devision of the superfamilies. His proposed classification included the subfamily Scythrinae in the Elachistidae.

During the last two decades phylogenetic analysis has become more detailed. Classification, previously based on imaginal, superficial structures has embodied a more diverse range of morphological characters, including those of immature stages, particularly in the systematic studies at suprageneric level. Common (1970) arrived, though without an exhaustive account of the group, at the view that the Scythrididae should be included in the large superfamily Gelechioidea which he placed between the Yponomeutoidea and the Copromorphoidea. In his later review of the evolution and classification of the Lepidoptera, Common's (1975) conclusions support his earlier point of view regarding the placement of the Scythrididae, though this is not expressed specifically.

The family ranking of Scythrididae has earlier been based on the following general characteristics (e.g. Meyrick, 1928):

(1) R_4 and R_5 in forewing stalked, R_4 to costa and R_5 to termen, (2) Rs and M_1 in hindwing parallel.

However, (1) is not applicable to all known genera (Hering, 1918; Powell, 1976) and (2) is inapplicable at least in the case of *Eretmocera medinella* (Stgr.) (Falkovitsh,

24

1981); moreover, neither trait is unique in the Gelechioidea. The affinities and status of the *Scythris* group of genera is obviously in need of further scrutiny.

The Nordic species may be recognized primarily by the 'drop-shaped' appearance of the moth in normal resting position, by the smooth-scaled (often glossy) lanceolate forewing and lanceolate hindwing, and, particularly, by the smooth and small head.

Genera within the Scythrididae

The cosmopolitan genus *Scythris* (s.l.) presently contains more than 370 described species of which at least 260 are confined to the Palearctic region. The extraordinary morphological variation of the group has deterred revision of the world fauna as has also the failure of early workers to refer to type specimens. Thus the generic name *Scythris* has, on the whole, been maintained in its broadest sense though more than 30 other genera have been described in the Scythrididae. However, none of these is involved amongst the species here dealt with, and it is beyond the scope of this work to give an exhaustive account of all genera but three genera of European interest may be mentioned. *Enolmis* Duponchel, 1845 *(Bryophaga* Ragonot, 1875) was established for the anomalous 'acanthella-group' (Leraut, 1980) and *Episcythris* Amsel, 1939 is used for a few other south European species. Species of *Enolmis* have pale forewings, a pterostigma in the forewing and divergent, asymmetrical genitalia. The genus *Eretmocera* Zeller, 1853 *(Staintonia* Staudinger, 1859), finally, comprises species mainly found in the tropical regions but has one European member *(E.medinella* (Stgr.)) (cf. Turati, 1924) and the generic characters are as follow: R_4 and R_5 coalescent, R_3 and M_1 from almost same point on cell and forewing red with black markings (Falkovitsh, 1981).

The previous arrangement of species in the Scythrididae originates from Zeller (1855) and has been used ever since. Species were classified according to their external appearance which has led to groupings being recognized which are now thought to be untenable. Here a new sequence of species is established, based mainly on the morphology of male and female genitalia. Zander (1905) pointed out two morphological groups but did not pay any attention to female genitalia. However, his grouping seems to be of certain importance and a further expansion of his ideas is expressed in the sequence of species adopted in this work.

The least transformed or reduced genitalia are considered to be the most primitive. Thus *S.obscurella* (Scop.) exhibits, on the whole, all components of the genitalia in an unmodified state. On the other hand, *S.restigerella* (Z.), for example, possesses very small and transformed male genitalia, completely deviating from the 'generalized' *Scythris* genitalia. *S.braschiella* (Hofm.) and *laminella* (Den. & Schiff.) seems to belong in an intermediate category.

A large species-group with members with more or less ordinary valvae is a provisional assemblage; it may in future be split up into further distinct groups.

Instead of describing new genera, Jäckh (1977, 1978b) has applied the 'species-group' concept which seems to be the most sensible approach with the state of our knowledge. As the phylogeny of this family is still little understood, the use of informal ranks is the best solution for the taxonomist, in order to bring a surveyable structure to the large family Scythrididae. A similar approach has been suggested by Johansson (1971) and Sattler & Tremewan (1978) for Nepticulidae and Coleophoridae respectively.

Zoogeography

The family Scythrididae is distributed worldwide but has its main centres around the Mediterranean, in Asia Minor and West Asia. Most species have a limited distribution and very few European species reach Siberia or Africa. The family is poorly represented in northern Europe but the number of species increase southwards with at least one hundred described species in the Mediterranean area. In central and southern Europe several *Scythris* occur in the alpine region. In northern Europe this is extremely rare; instead, species are found in arid areas near the coast. Thus the southern parts of Norway, Finland and Sweden have most species. In Denmark, conditions are somewhat different and a clear picture does not emerge at present. However, there seems to be a concentration of species in East Jutland and eastern North Jutland.

Eleven species are known from Denmark, 15 from Sweden, 8 from Norway and 11 from Finland. A total of 18 species is recorded from the Nordic countries. Only one species is exclusively confined to this region *(S.fuscopterella* Bengts.) but one record from France may belong to this species. *S.braschiella* (Hofm.) has not been collected outside two districts in N Germany and might be endemic there since its environmental requirements are apparently very specific. Disjunct distribution is exhibited by, for example, *crypta* Hann., known from southern Sweden, Italy and Macedonia. Only one Holarctic species has so far been recognized *(S.limbella* (F.)), but it has possibly been introduced into North America by man. *S.noricella* (Z.) might be Holarctic but there is no record from North America to confirm this. However, *noricella* is recorded from Greenland as ssp. *longifoliella* Wolff. *S.ericivorella* (Rag.) has a pronounced atlantic distribution, *S.crypta* Hann. belongs to the mediterranean group while the rest of the Nordic species may be regarded as boreal or boreo-montane species.

Technical remarks

Wing preparations were made using the method described by Traugott-Olsen & Nielsen (1977); wings were mounted in canada balsam after staining in Chlorazol Black.

Genitalia were macerated in boiling 10% caustic potash (KOH) for 2-5 minutes until the abdomen appeared transparent; it was then washed in several changes of destilled water. The genitalia passed through increasing concentrations of ethyl alcohol and, if necessary, stained using Chlorazol Black. After some 5-15 minutes in 99% alcohol with careful agitation, the genitalia were transferred to xylene and then mounted in canada balsam.

Although the genitalia were usually removed from the rest of the abdomen before they were drawn, no further dissection was carried out. Instead they were studied and, for later identification of the different sclerites, sketched in a lateral position after maceration and washing.

The illustration of adults is laborious and troublesome since most *Scythris* species have unmarked forewings but with reflected colours which may appear different depending on the colour temperature of the light-source and the angle of illumination. In daylight the reflected colour of the moths may be somewhat different and usually they look a bit darker to the naked eye than in a stereomicroscope, depending on the light-intensity.

Genus *Scythris* Hübner

Scythris Hübner, (1825): 414.
Type-species: *Tinea chenopodiella* Hübner, (1813), = *(Tinea) limbella* Fabricius, 1775.
Galanthia Hübner, (1825): 417.
Type-species: *Galanthia extensella* Hübner, (1825), = *(Phalaena) obscurella* Scopoli, 1763.
Butalis Treitschke, 1833: 108, nec Boie 1826.
Type-species: *Tinea knochella* Fabricius, 1794.

Keys to species of *Scythris*

Species marked with an asterisk are those found in Fennoscandia and/or Denmark.

A. Key based on external characters, best applicable to males.

(N.B. Species often cannot be safely identified by external characters and in cases of doubt genitalia dissection must be resorted to.)

1	Forewing without markings or pale scales	2
–	Forewing with markings or pale scales	23
2 (1)	Forewing with yellowish or greenish reflection, more or less glossy	3
–	Forewing with bluish or purplish reflection or without gloss	14
3 (2)	Anal brush in male with curly tufts, pale ochreous	4
–	Anal brush with straight, usually fuscous tufts	5
4 (3)	Abdomen dorsally dark, fuscous (Fig. 60) .... 33. *grandipennis* (Hw.)	
–	Abdomen dorsally pale, ochreous (Fig. 61) .... 34. *ericetella* (Hein.)	
5 (3)	Anal brush in male distinctly trifid, central portion large	6
–	Anal brush compact, at most with very weak tripartition	10
6 (5)	Hindwing about half as broad as forewing; wing expanse less than 13 mm; forewing with yellowish fuscous reflection (Fig. 40) .... 15. *palustris* (Z.)*	
–	Hindwing more than two thirds as broad as forewing; wing expanse more than 13 mm	7
7 (6)	Forewing without greenish reflection (Fig. 33) .... 8. *scopolella* (L.)	
–	Forewing with more or less greenish reflection (three spe-	

cies, generally impossible to differ from the external
appearence) 8
8 (7)   Forewing with yellowish green reflection; valva with
median angle dorsally; aedeagus long and slender (Figs.
36, 73) .... 11. *seliniella* (Z.)
   –   Forewing with greyish green reflection 9
9 (8)   Valva straight, rather slender (Fig. 74) .... 12. *subseliniella* (Hein.)
   –   Valva with median angle dorsally; aedeagus short (Fig.
72) .... 10. *clavella* (Z).
10 (5)   Abdomen conspicuously thick, dark-scaled, terminal
segments with central depression 11
   –   Abdomen slender, if stout then with many yellowish scales 12
11 (10)   Wing expanse less than 12 mm (Fig. 53) .... 27. *crassiuscula* (HS.)
   –   Wing expanse more than 12 mm (Fig. 62) .... 35. *fallacella* (Schl.)
12 (10)   Wing expanse usually more than 19 mm; hindwing as
broad as forewing (Fig. 23) .... 1. *obscurella* (Scop.)*
   –   Wing expanse usually less than 18 mm; hindwing at most
0.8 times as broad as forewing 13
13 (12)   Forewing dark green; wing expanse usually more than
17 mm (fig. 39) .... 14. *productella* (Z.)*
   –   Forewing paler green, with faint yellowish or orange
reflection; wing expanse generally less than 17 mm (Fig.
59) .... 32. *fuscoaenea* (Hw.)*
14 (2)   Abdomen mostly bright yellow-orange (Fig. 38) 13. *sinensis* (Feld. & Rog.)
   –   Abdomen dorsally fuscous 15
15 (14)   Hind tarsi externally whitish (Fig. 28) .... 5. *bifissella* (Hofm.)
   –   Hind tarsi externally not whitish 16
16 (15)   Forewing greyish with yellow or brown reflection; labial
palpus small, straight (Fig. 34) .... 9. *paullella* (HS.)
   –   Forewing more or less dark brown, usually not with yellow reflection .. 17
17 (16)   Forewing with distinct purplish reflection, at least in apical area 18
   –   Forewing without distinct purplish reflection 20
18 (17)   Hindwing narrow, only about half as broad as forewing;
labial palpus small, straight (Fig. 52) ...... 26. *laminella* (Den. & Schiff.)*
   –   Hindwing almost as broad as forewing; labial palpus curved 19
19 (18)   Scales in forewing not with pale base; uncus ending in a
point (Figs. 49, 89) .... 23. *disparella* (Tgstr.)*
   –   Scales in forewing with pale base; uncus furcate or with
concave end (Figs. 48, 84-88) .... 22. *picaepennis* (Hw.)*
20 (17)   Forewing olive brown, glossy (Fig. 47) .... 21. *tributella* (Z.)
   –   Forewing dark brown, rather dull 21
21 (20)   Hindwing almost as broad as forewing (Fig. 54) .. 28. *ericivorella* (Rag.)*
   –   Hindwing about two thirds as broad as forewing; first

37 (31)	Forewing with two distinct spots, one in fold and one at cell end; ground colour dark violet-brown (Fig. 41) .. 16. *muelleri* (Mann)
–	Forewing with other markings 38
38 (37)	Forewing with three distinct spots and a basal oblique dash (Fig. 32) .... 8. *scopolella* (L.)
–	Forewing with other markings 39
39 (38)	Wing expanse more than 16 mm; ground colour greenish, glossy; a double spot near base and an apical dash, all yellowish (Fig. 24) .... 2. *cuspidella* (Den. & Schiff.)
–	Wing expanse less than 12 mm; forewing otherwise 40
40 (39)	Forewing brown or dark grey with two pale brown spots, one smaller posterior to fold and one larger at cell end (Fig. 51) .... 25. *braschiella* (Hofm.)
–	Forewing otherwise 41
41 (40)	Forewing dark grey, with several white markings, at least two elongate spots in fold; abdomen rather slender, in female bluish grey; second segment of labial palpus in full entirely white (Fig. 45) .... 19. *empetrella* Karsh. & Niel.*
–	Forewing dark brown, at most with two small, white spots in fold; second segment of labial palpus whitish only in basal half 42
42 (41)	Pale, ochreous scales usually present in apical area; hindwing with curved hind margin (Fig. 46) .... 20. *siccella* (Z.)*
–	Pale, ochreous scales absent in forewing; hindwing with straight hind margin (Figs. 55, 56) .... 29. *crypta* Hann.*

B. Key based on male genitalia

1	Valvae symmetrical or almost symmetrical, occasionally hardly identifiable 2
–	Valvae distinctly asymmetrical (Figs. 70, 79, 82, 90, 91) 31
2 (1)	Valva cygnate, near base with field of close-set, short setae along lower margin (Fig. 83) .... 21. *tributella* (Z.)
–	Valva of different shape and with different vestiture 3
3 (2)	Genitalia compact, often embraced in full by segment VIII; valva a short lobe or a small, occasionally hardly identifiable fold 4
–	Genitalia not particularly compact; valva long 14
4 (3)	Valva at most developed as a very small fold with bristles (Figs. 93, 94, 96-101) 5
–	Valva a triangular or rounded lobe (Figs. 84, 88, 89, 95) 12
5 (4)	Aedeagus diminutive, more or less bottle-shaped 6
–	Aedeagus long and slender, more or less curved 10
6 (5)	Tegumen with bent, pointed processes (Fig. 101) ...... 35. *fallacella* (Schl.)

C. Key based on female genitalia

	extension (Fig. 106) .... 5. *bifissella* (Hofm.)
-	Lamella postvaginalis with anterior, sclerotized extension (Fig. 110) .... 9. *paullella* (HS.)
5 (1)	Lamella postvaginalis more or less triangular, sometimes with an anterior swelling or emargination 6
-	Lamella postvaginalis of different shape 16
6 (5)	Ductus bursae distinctly sclerotized, widened posteriorly and anteriorly (Fig. 135) .... 34. *ericetella* (Hein.)
-	Ductus bursae not sclerotized 7
7 (6)	Lamella postvaginalis drop-shaped, with central unmelanized 'window' (Fig. 125) .... 24. *fuscopterella* Bengts.*
-	Lamella postvaginalis of different shape 8
8 (7)	Lamella postvaginalis with lateral processes at midlength and with unmelanized 'window' (Fig. 121) .... 20. *siccella* (Z.)*
-	Lamella postvaginalis different 9
9 (8)	Apophyses anteriores much shorter than width of segment VIII 10
-	Apophyses anteriores not shorter than width of segment VIII 11
10 (9)	Apophyses anteriores straight (Fig. 124) .... 23. *disparella* (Tgstr.)*
-	Apophyses anteriores curved (Fig. 123) .... 22. *picaepennis* (Hw.)*
11 (9)	Lamella postvaginalis much shorter than segment VIII 12
-	Lamella postvaginalis subequal to length of segment VIII 13
12 (11)	Lamella postvaginalis broad, extensively unsclerotized medially. Apophyses posteriores considerably longer than ap.anteriores (Fig. 127) .... 26. *laminella* (Den. & Schiff.)*
-	Lamella postvaginalis narrower and more fully sclerotized. The two pairs of apophyses of subequal length (Fig. 126) .... 25. *braschiella* (Hofm.)
13 (11)	Lamella postvaginalis with paired anterior sclerites (Fig. 129) .... 28. *ericivorella* (Rag.)*
-	Lamella postvaginalis without paired anterior sclerites 14
14 (13)	Lamella postvaginalis considerably stronger sclerotized along lateral/posterior margins than in central and anterior parts, reminescent of a boomerang (Fig. 128) .. 27. *crassiuscula* (HS.)
-	Lamella postvaginalis with almost uniform sclerotization except for a minute, posterior slit 15
15 (14)	Sternum VII concave posteriorly (Fig. 134) .... 33. *grandipennis* (Hw.)
-	Sternum VII convex posteriorly (Fig. 122) .... 21. *tributella* (Z.)
16 (5)	Lamella antevaginalis a long plate, its anchor-shaped anterior part fitting into a broad concavity in sternum VII (Fig. 119) .... 18. *noricella* (Z.)*
-	Lamella antevaginalis and sternum VII of different shape 17
17 (16)	Lamella antevaginalis fingershaped (Fig. 118) .... 17. *inspersella* (Hb.)*
-	Lamella antevaginalis different 18

33

34

28 (27)   Apophyses posteriores more than 1.6 mm (Fig. 105) ..... 4. *cicadella* (Z.)*
\-   Apophyses posteriores less than 1.5 mm (Fig. 104) ... 3. *potentillella* (Z.)*
29 (27)   Lamella postvaginalis trapezoid, extensively unsclerotized medially, considerably shorter than segment VIII (Fig. 127) .... 26. *laminella* (Den. & Schiff.)*
\-   Lamella postvaginalis differently developed 30
30 (29)   Sternum VII posteriorly with a single, paramedial process and a deep medial incision (Fig. 116) .... 15. *palustris* (Z.)*
\-   Sternum VII of different shape 31
31 (30)   Lamella postvaginalis with asymmetrical posterolateral humps (Fig. 120) .... 19. *empetrella* Karsh. & Niel.*
\-   Lamella postvaginalis of different shape 32
32 (31)   Lamella postvaginalis arrow-shaped (Fig. 121) .... 20. *siccella* (Z.)*
\-   Lamella postvaginalis not arrow-shaped 33
33 (32)   Lamella postvaginalis drop-shaped, with central 'window' (Fig. 125) .... 24. *fuscopterella* Bengts.*
\-   Lamella postvaginalis of different shape 34
34 (33)   Tergum VII very extended laterally, laterally rounded (Fig. 136) .... 35. *fallacella* (Schl.)
\-   Tergum VII of different shape 35
35 (34)   Lamella postvaginalis complex, almost trapezoid, with anterior, oval 'window' (Fig. 108) .... 7. *knochella* (F.)*
\-   Lamella postvaginalis of different shape 36
36 (35)   Lamella postvaginalis boomerang-shaped (Fig. 128) 27. *crassiuscula* (HS.)
\-   Lamella postvaginalis otherwise 37
37 (36)   Lamella postvaginalis consists of two large, curved sclerotizations, more or less extensively fused in mid-line (Fig. 117) .... 16. *muelleri* (Mann)
\-   Lamella postvaginalis different 38
38 (37)   Sternum VII with posterior emargination (Fig. 107) ...... 6. *limbella* (F.)*
\-   Sternum VII without posterior emargination 39
39 (38)   Sternum VII with a postero-median hump (Fig. 130) ... 29. *crypta* Hann.*
\-   Sternum VII evenly curved posteriorly (Fig. 115) ..... 14. *productella* (Z.)*

1. *Scythris obscurella* (Scopoli, 1763)*
Figs. 23, 63, 102.

Phalaena obscurella Scopoli, 1763: 252.
Tinea esperella Hübner, 1799: 255.
?*Galanthia extensella* Hübner, (1826): 417.

17-21 mm. Head, labial palpi, antennae and thorax dark grey with more or less greenish (male) or brownish (female) tone. Collar and tegulae usually with more markedly green or brown gloss. Forewing dark bronzy, in female purple-tinged in apical area. Hindwing fuscous, sometimes with faint violet tinge; scales 2-3 times as long as broad. Cilia of forewing and hindwing fuscous. Upperside of male abdomen glossy dark fuscous. Underside of same colour as upperside but anal brush off-white posteriorly. Female abdomen a dusky yellowish dorsally but three posterior segments dark brownish. Underside dark fuscous with two penultimate segments ivory. Specimens with aberrant markings have been recorded from different areas. Such specimens have caused some confusion in the past (Zeller, 1885); thus v. *flavidorsella* (Rbl.) has an entirely yellow abdomen.

Diagnosis. *S.obscurella* (Scop.) is easily recognizable by its size and colour and ought not to be confused with the other glossy northern species. Only *productella* (Z.) may reach the same size but its forewing is narrower, darker, bottle-green, without yellowish gloss; the hindwing of *obscurella* is almost twice as broad as in *productella*.

Male genitalia (Fig. 63). Valva rather slender, evenly curved and pointed. Uncus two-pronged, not pointed, with a rather broad sinuation. Gnathos large, tapered and pointed. Aedeagus thin, straight at base, curved and tapered in apical half. Sternum VIII triangular, posteriorly divided in two pointed and long projections.

Female genitalia (Fig. 102). Lamella postvaginalis a ring-shaped sclerotization, usually incomplete anteriorly. Posterior part of ductus bursae weakly sclerotized, slender. Very similar to *cuspidella* (Den. & Schiff.) in which the lamella postvaginalis is a complete, strongly sclerotized ring.

Distribution. Recorded from three districts in southeastern Finland. Found also in Latvia and Estonia (Šulcs, 1981). Known from C and E Europe to Asia Minor and C Asia. Mostly in mountains; in Finland and Latvia in meadows.

Biology. Spuler (1910) and Hering (1932) state that the larva may be found on Leguminosae during spring. Adults fly in June in Finland but usually later in the south. There is a single record of a specimen having been collected at light (Šulcs, 1981).

Note. *S.lampyrella* (Constant, 1865) is here considered to be conspecific with *thomanni* (Müller-Rutz, 1914) (Passerin d'Entrèves, 1977). Jäckh (in litt.), on the other hand, regards *lampyrella* to be a form of *obscurella*.

S.obscurella belongs to one of the most critical groups of *Scythris*. At least 8 species occur in the Alps and they are very difficult to keep apart. Externally the species of this '*obscurella*-group' vary in size, coloration and markings, and the genitalia of the males are not distinctive. Only the female genitalia may, in most cases, offer means for determination.

2. *Scythris cuspidella* ((Denis & Schiffermüller), 1775)
Figs. 24, 64, 103.

Tinea cuspidella Denis & Schiffermüller, 1775: 140.
Tinea bifariella Hübner, 1813: figs. 385, 386.

16-20 mm. Head dark brown with some paler scales above eyes. Labial palpi dark brown with whitish scales laterally. Antennae, tegulae and thorax dark brown. Collar with contrasting scales, whitish to dark brown. Forewing glossy olive-brown with, near base, a whitish or cream fascia, extending from costa to fold and obliquely posteriorly-directed; fascia sometimes interrupted at Sc or even almost absent. Sometimes a few whitish scales at end of cell between apex and tornus. Apex with a yellowish streak, becoming obsolete towards costa. The variation in forewing pattern is considerable; the extension of the cream spots may vary a great deal and they may be partly or wholly lost. The wing-shape may vary (Heinemann, 1877). Hindwing 0.8-0.9 times as broad as forewing, dark brown. Cilia of hindwing and forewing fuscous. Male abdomen very dark fuscous; anal tuft somewhat paler. Female abdomen gradually tapering with protruded ovipositor; dorsally fuscous or dark brown; ventrally paler, brown ochreous, with cream subventral, more or less fused, blotches on the two penultimate segments.

Diagnosis. Easily recognizable by the cream spots in the forewing. More uniformly coloured forms may resemble *obscurella* (Scop.) or other large, glossy species but the dark olive-green colour and comparatively broad forewing is typical for *cuspidella*.

Male genitalia (Fig. 64). Valva curved, abruptly pointed, outer margin with two indistinct, rounded bends, one in middle and one near tip. Gnathos stout, thorn-like. Uncus horseshoe-shaped. Aedeagus slender, tapered. Sternum VIII triangular, laterally concave and with shallow bifurcation at posterior end.

Female genitalia (Fig. 103). Very similar to *obscurella* (Scop.); lamella postvaginalis ring-shaped, thicker posteriorly and with small indentation. Two symmetrical, paramedian, melanized patches posterior to the ring.

Distribution. Not recorded from Denmark or Fennoscandia. Reported from FRG, GDR, Poland, Austria, Czechoslovakia, Hungary, Rumania, Bulgaria, Italy, Yugoslavia, Greece and the Caucasus. Also Ural Mts (Petersen, 1923). The nearest records are from Harz and Thuringia.

Biology. The larva is unknown. Falkovitsh (1981) states *Thymus* as a possible host plant. Adults appear from the end of June to mid-August in meadows in deciduous forests or in mountain areas.

3. *Scythris potentillella* (Zeller, 1847)*
Figs. 4, 25, 65, 104.

Butalis potentillella Zeller, 1847: 832.
Butalis potentillae Zeller, 1855: 202.
Scythris albiapex M. Hering, 1924: 79.

10-13 mm. Head, labial palpi, antennae, collar, tegulae and thorax dark brown with very faint bluish tinge. Forewing dark brown with a hardly discernible bluish or purplish reflection and often with paler scales of white, grey and ochreous shades. Occasionally such scales appear in great numbers, making the apical area considerably lighter and sometimes forming a more or less distinct streak in the fold, such specimens appearing similar to *cicadella* (Z.). Hindwing 0.7-0.8 times as broad as forewing, brown, darkened towards apex. Cilia of both forewing and hindwing fuscous. Male abdomen dark fuscous, ventrally with som pale scales. Female abdomen dark fuscous with a few pale scales on ventral side. Last segment with dirty yellowish scales ventrally.

Diagnosis. *S.potentillella* (Z.) may be confused with *picaepennis* (Hw.) but it normally has white and not only ochreous scales in the forewing; *potentillella* is also less glossy, especially the hindwing. *S.disparella* (Tgstr.) has no pale scales on the forewing. *S.cicadella* (Z.) has blotches of ochreous scales on the forewing. Other comparable species are considerably more glossy.

Male genitalia (Fig. 65). Similar to *cicadella* (Z.) (Fig. 66) but differing as follows. Valva slightly curved in apical half, rapidly tapering to a point which is directed straight towards the tip of the other valva. Gnathos tapered to a slender, straight and flattened point. Uncus with rounded lateral base. Aedeagus regularly tapering, weakly curved. Tergum VIII with shallow, posterior emargination.

Female genitalia (Fig. 104). Very similar to *cicadella* (Z.). Lamella postvaginalis with a broader, arched sclerotization, posteriorly having a cygnate structure, which is normally folded back anteriorly in preparation. Bridge connecting apophyses anteriores and their posterior extension U-shaped, this armature often shorter than broad. Apophyses posteriores 1.2-1.5 mm and apophyses anteriores 0.5-0.8 mm (cf. *cicadella*). Sternum VII rectangular, posteriorly and anteriorly slightly concave, often broader than long.

Distribution. In Denmark known from Jutland, Zealand and Bornholm; in Sweden in the southern half and along the coast to Nb; Norway: On; in Finland found up to latitude 65°. Great Britain (Suffolk) and C Europe.

Biology. Larva dark brown with pale dorsal line; head and prothoracic shield black (Spuler, 1910). It feeds on *Rumex acetosella* L. and lives in a web tube along the stem and on the ground. The phenology of the larva is insufficiantly known in northern Europe. Adults have been collected over a long period, from the end of May to the end of August, and there are probably two more or less well-defined broods. *S.potentillella* is usually met with in sandy places, on open moraine hills, slopes or hummocks. It is occasionally found in alpine environments, just above the tree limit.

Note. In the original description Zeller (1847) spelled the species name 'potentillella' but later Zeller (1855) changed the spelling to 'potentillae'. The latter name has been used ever since until Leraut (1980) reintroduced the original name.

S.albiapex M. Hering was synonymized with *potentillella* by Hackman (1945).

4. *Scythris cicadella* (Zeller, 1839)*
Figs. 5, 26, 27, 66, 105.

Butalis cicadella Zeller, 1839: 193.

10-12 mm. Head dark brown, scape with white scales. Antennae dark fuscous. Basal segment of labial palpus white; second segment richly provided with white scales; terminal segment fuscous with fewer white scales. Thorax, collar and tegulae dark brown, the latter with some pale scales at tip. Forewing pattern very variable. Ground colour dark brown, most often with distinct white markings. Darker forms almost completely lack white scales except in the fold. Normal individuals have the following whitish markings, sometimes mixed with ochreous scales: streak in fold, small dash at dorsum near base, blotch before tornus and a smaller one beyond this, often connected with a blotch above tornus in middle of forewing and, finally, a patch at apex. Further white scales are often scattered in other areas of forewing, especially near costa. Hindwing 0.7-0.8 times as broad as forewing, brownish. Cilia fuscous in both forewing and hindwing. Male and female abdomen dorsally fuscous, ventrally brownish with many whitish scales which are denser posteriorly. Anal tuft almost completely composed of whitish scales in lightly marked specimens, otherwise tuft most often pale fuscous, at least dorsally.

Diagnosis. The white and ochreous markings on the forewing are characteristic of *cicadella*. Only *potentillella* (Z.) may exceptionally be confused with *cicadella* but *potentillella* usually exhibits at most a narrow, white streak in the fold and scattered white scales; in such cases the genitalia will provide a reliable identification.

Male genitalia (Fig. 66). Very similar to *potentillella* (Z.) but valva more gradually tapering and the direction of the tips is more posterior. Gnathos evenly tapered, slightly curved; uncus with pointed, lateral base; tergum VIII with rather deep, posterior emargination.

Female genitalia (Fig. 105). Very much resembling *potentillella* (Z.). Lamella postvaginalis U-shaped with cygnate sclerotizations posteriorly. Tergum VIII armature appearing nearly concentric with lamella postvaginalis, larger than in *potentillella;* apophyses posteriores 1.6-1.9 mm and apophyses anteriores 0.8-0.9 mm (cf. *potentillella).*

Distribution. In Denmark in five western districts; in Sweden north to Ög. Not in Norway or Finland. Otherwise reported from SE England, Lithuania and several parts of C Europe.

Biology. According to Benander (1965) the larva is greenish yellow with a white dorsal line and dark brown subdorsal and lateral lines; head and prothoracic shield blackish brown with yellow dots; anal plate yellowish green; abdominal prolegs dark brown. It feeds on *Scleranthus annuus* L. and *S.perennis* L. in June, making a long silken tube beneath the plant and amongst moss. The web is often mixed with sand grains. Spuler (1910) states that the larva has a yellow-brown ground colour and a similarly coloured head. The larva may, according to observations made by the author, also feed on some other plant as *cicadella* is found in numbers of places where *Scleranthus* does not grow (Öland, Sweden).

Adults appear from mid-June through July. They are usually active only in bright sunshine, when they jump or fly over low vegetation in sandy places.

Note. The characteristics of the male genitalia were first pointed out by Wolff (1959) who figured the genitalia of Zeller's original material, kept in the British Museum (N.H.).

5. *Scythris bifissella* (Hofmann, 1889)
Figs. 28, 67, 106.

Butalis bifissella Hofmann, 1889: 107.

9-12 mm. Head glossy copper-brown; labial palpi with many whitish scales on a fuscous ground; antennae dark brownish, scape ventrally whitish, basal half of antennae ventrally with a few whitish scales; collar whitish; thorax and tegulae pale brownish. Forewing brownish, infuscated, olive-tinged, with some paler scales in apical area and, sometimes, around fold. Cilia brownish, infuscated, some pale scales at base, especially at termen; costal fringes with whitish tips. Hindwing 0.8 times as broad as forewing, dark brown. Cilia dark brown. Hind tibia whitish; hind tarsi ventrally white. Male abdomen dark fuscous dorsally; ventral side paler, with ochreous brown and whitish scales; anal tuft long, fuscous, in dorsal view divided into two brushes; tuft ventrally yellowish in middle. Female abdomen according to Hofmann (1889) swollen, tapering posteriorly and broadly truncated; blackish dorsally, glossy; ventrally paler due to numerous white and yellowish scales, especially at posterior margins of segments and on posterior segments; papillae anales protruding.

Diagnosis. Similar to *potentillella* (Z.) but smaller, hind tarsi white ventrally and anal brush different, longer and with a yellowish ventromedial tuft.

Male genitalia (Fig. 67). Valva evenly curved, distally broadening, terminally rounded; uncus very deeply furcate; gnathos very long, tapered, pointed, with wide base; aedeagus tubiform, tapered, slightly curved; tergum VIII (not figured) triangular, furcate posteriorly. Genitalia reminiscent of *S. flavidella* Preiss. (from S Europe).

Female genitalia (Fig. 106). Lamella postvaginalis circular, with lateral, sclerotized processes, posteriorly with a weak sclerotization; lamella antevaginalis weak, rectangular or oval.

Distribution. Not in Denmark or Fennoscandia. Recorded from C Europe; nearest in central Germany (Thuringia).

Biology. See Rapp (1936). Larva pale with pale dorsal line and with fine, reddish brown subdorsal and lateral lines sometimes irregularly broken; head pale with small, reddish dots; prothoracic shield formed by two large, rounded blotches. It feeds in June on *Silene otites* (L.), living in concealed silken galleries below the basal leaves and among the dead parts and fresh leaves of the plant. Adults fly from mid-June to the end of July.

6. *Scythris limbella* (Fabricius, 1775)*
Figs. 6, 17, 29, 30, 68, 107.

Tinea limbella Fabricius, 1775: 660.
Tinea variella Denis & Schiffermüller, 1775: 140.
Tinea quadriguttella Thunberg, 1794: 87.
Tinea tristella Hübner, 1796: 218.

Tinea chenopodiella Hübner, 1813: 320.
Glyphipteryx cylindrella Stephens, 1834: 280.

14-16 mm. Head, labial palpi, collar, tegulae and thorax dark brown mixed with dirty cream scales to a varying extent. Antennae dark brown. Ground colour of forewing dark olive brown with the following cream markings: a plical streak to middle of wing, two subplical dashes connected with streak in fold, an apical blotch and some basal scales at dorsum. In darker specimens the light markings may be reduced to only one or two small patches (f. *obscura* (Stgr.)). Cilia brown, brighter towards apex and in most specimens with whitish, subapical dash. Hindwing 0.8 times as broad as forewing, brown, darker towards apex, base partly without scales. Cilia as in forewing. *S.limbella* f. *obscura* (Staudinger, 1870) (Fig. 30) has a very dark forewing with reduced pale markings. This form has been found at least in Finland (Hackman, 1945) and in Sweden (by the author). Male abdomen with upper side brown with somewhat paler scales at posterior edge of segments. Underside wholly whitish with very small admixture of brownish scales. Female abdomen dark brown with yellow scales at posterior edges of segments. In dark specimens these scales may be missing but the two penultimate segments at least exhibit yellow scales. Underside whitish.

Diagnosis. *S.limbella* may be confused with *knochella* (F.) but has more extensive and less distinct markings in the forewing; the ground colour is dark brown without a violet or reddish tone.

Male genitalia (Fig. 17, 68). Valva curved, of uniform width from middle, apex tapering, with rounded tip; uncus U-shaped, slightly asymmetrical; gnathos large, S-shaped, tapered and pointed; aedeagus short, almost straight, cylindrical, thickened towards base and with smaller subapical flexure; sternum VIII deeply cleft posteriorly and with apically rounded tips.

Female genitalia (Fig. 107). Lamella postvaginalis characteristically shaped, posteriorly with a bilobed structure which terminates in a weak sclerotization with concave lateral margins; sternum VII distally narrowing, with incurved apical margin.

Distribution. Recorded from five districts in Denmark; in Sweden north to Nb., mostly in the eastern part; from six southeastern provinces in Norway; Finland: in several southern provinces, also ObS. Further distributed over the whole of C and S Europe to Asia Minor and Turkestan. Sheppard (1974) recorded this species from the NE USA and Canada.

Biology. Spuler (1910) and Meyrick (1928) state the larva to be grey with olive green tone; dorsal and subdorsal lines fine, cream; head grey with black marbling; prothoracic shield blackish; anal plate with two large, black spots. Larvae live gregariously in a silken web on leaves, in flowers and in buds of *Chenopodium* and *Atriplex* in (April-)May to August, in Denmark probably in two overlapping broods. The phenology of *limbella* in N Europe is little-known and there are indications that individuals aestivate. Schütze (1931) also expessed uncertainty as to the number of broods. Adults fly from the end of May to September but are most frequently found at the end of July and in early August. The moths may often be seen on walls, poles and lattices.

Note. *S.limbelloides* Jäckh was described from Spain and this species could easily be confused with *limbella*. Specific differences can only be found in the genitalia (Jäckh, 1978a).

7. *Scythris knochella* (Fabricius, 1794)*
Figs. 7, 31, 69, 108.

Tinea knochella Fabricius, 1794: 318.

12-14 mm. Head, labial palpi, antennae, collar, tegulae and thorax brown. Forewing deep brown with faint purplish tinge, slightly glossy. A distinct pale yellowish streak of constant width in the fold, this streak ending immediately beyond middle of forewing. At end of cell a yellowish spot with straight or, more often, concave outer margin. Cilia dark fuscous. Hindwing 0.7 times as broad as forewing, brown. Cilia dark fuscous with paler base. Male abdomen uniformly fuscous; anal tuft ventrally ivory, obscuring terminal bifid projections of sternum VIII. Female abdomen with segments II-IV blackish brown dorsally, noticeably darker than other segments which are brown. Terminal segment with posterior crown of yellow scales which may be missing dorsally. Underside ivory, anteriorly darkened by admixture of brownish scales.

Diagnosis. *S.knochella* (F.) may be confused with *S.limbella* (F.) and *cicadella* (Z.) which, however, have more extensive and less distinct markings; the ground colour is without a violet tinge in both *limbella* and *cicadella*.

Male genitalia (Fig. 69). Valva skittle-shaped, terminally slightly bent; gnathos pointed, with weak curvature, base a complicated sclerotization with many minute parallel callosities; uncus resembling a horseshoe with broad shanks, each carrying a rounded appendage; aedeagus of moderate length, bent and tapered in terminal third; sternum VIII deeply bifid terminally.

Female genitalia (Fig. 108). Ostium with strong rim; lamella postvaginalis with narrow border anteriorly, posteriorly with two abruptly broadening and compact arms.

Distribution. In Denmark in one site in northwestern Zealand (Bjergsted) (Pallesen & Palm, 1974); in Sweden only in Sk. where it occurs commonly around Lake Krankesjön (Samuelsson, 1976); not in Norway or Finland. Reported from GDR, FRG, Czechoslovakia, France and Spain (Jäckh, 1978b). Also from Belgium (Coenen, in litt.). In southern Europe *S.knochella* seems to be superseded by *S.punctivittella* (Costa), *aspromontis* Jäckh and other similarly marked *Scythris*.

Biology. The immature stages are incompletely known. Spuler (1910) described the larva as greenish white with narrow, reddish yellow dorsal and subdorsal lines; head and prothoracic shield yellowish with darker spots. Observations by several collectors suggest that the larva feeds on *Cerastium arvense* L. and *C.semidecandrum* L.. Jäckh (1978b) records finding a larva on *Thymus*. According to Schütze (1931) the larva appears in May and June, living in a web at the root of the host plant. The web is extended along the stem and onto the ground. Adults fly from early June to mid-August on dry and open ground rich in herbs.

42

Note. The single specimen (without abdomen) of *knochella* in the Fabrician collection (Zoological Museum, University of Kiel, deposited on permanent loan in the Zoological Museum, University of Copenhagen) has been designated as lectotype by O. Karsholt and is cited here for the first time: 'Knochella' (in presumed Fabrician handwriting) – 'Lectotype, *Tinea knochella* Fabricius, 1794. O. Karsholt design. 1984'.

8. *Scythris scopolella* (Linnaeus, 1767)
Figs. 32, 33, 70, 109.

Phalaena (Tinea) scopolella Linnaeus, 1767: 896.
Butalis triguttella Zeller, 1839: 193.
Lita triguttella Duponchel, 1839: 332.
Butalis heleniella Millière, 1876: 361.
Scythris gredosella Schmidt, 1941: 38.

13-14 mm. The species appears in two forms, one with markings and one without. Head, labial palpi, antennae, collar, tegulae, thorax and forewing more or less dark brown with an olive tone; unicolorous specimens usually paler with slight yellowish reflection; marked specimens generally dark brown, forewing paler basally and with following whitish markings: oblong dorsal patch at base, square blotches in middle of fold and at tornus, and rounded spot at apex. Cilia dark fuscous. Hindwing as broad as forewing, dark greyish brown with faint violet tinge. Cilia dark fuscous. Male abdomen dark fuscous dorsally, greyish ochreous ventrally; anal tuft trifid, ochreous brown. Female abdomen dorsally brown with faint greyish shade; ventrally mostly whitish, more or less mixed with dark ochreous scales; two penultimate segments completely white ventrally, the white colour extended dorsolaterally, and visible from above.

Diagnosis. Unicolorous specimens resemble several other species; *subseliniella* (Hein.), *seliniella* (Z.) and *clavella* (Z.) all have a longer labial palpus (three times the diameter of the eye, in *scopolella* less than twice); *ericetella* (Hein.) has a curled anal tuft in the male; in the female the white underside colour does not extend to the subventral area.

Male genitalia (Fig. 70). Valvae slightly asymmetrical, furcate, with rounded ventral process; uncus a small bowed hood; gnathos large, pointed, basally broad, with granular patch dorsally; aedeagus bent in middle, tapered, a long spine above base; sternum VIII furcate, with rather long branches.

Female genitalia (Fig. 109). Lamella postvaginalis elongate, posteriorly broadened and rounded; antrum distinctly sclerotized, funnel-shaped; ductus bursae with membrane progressively thinner anteriorly.

Distribution. Not in Denmark or Fennoscandia. Central and southern FRG, GDR, Belgium, France, Switzerland, Austria, Czechoslovakia and Italy. A female specimen from Poland (Szczecin) in the Zoological Museum, Copenhagen; not earlier recorded from Poland.

Biology. Larva brown, with paler lines; on moss (Spuler, 1910). Schütze (1931) related observations made by Steudel (who found the larva at early May om *Tortula muralis* (L.) in fine silken tubes), Disque (who found the larva at the end of June under *Helianthemum, Hypericum* and moss on walls), and Reutti (who suggested *Sedum album* L. as the host plant). Adults occur from the end of May to the end of July on dry slopes and around mossy walls.

Note. As discussed by Robinson & Nielsen (1983) the correct authorship of the name *scopolella* is Linnaeus, 1767, not Hübner, 1799 (or 1796). Passerin d'Entrèves (1979) gave further synonymies of this species and checked the type of *gredosella* Schmidt.

9. *Scythris paullella* (Herrich-Schäffer, 1855)
Figs. 34, 71, 110.

Oecophora paullella Herrich-Schäffer, 1855: 270.

11-12 mm. Head, labial palpi, antennae, tegulae and thorax olivaceous fuscous, collar slightly paler; labial palpi straight, descending. Forewing grey with yellowish or brownish tone, rather dull. Hindwing 0.7-0.8 times as broad as forewing, brownish, somewhat paler than forewing. Cilia of forewing and hindwing brownish. Male abdomen rather slender but short, pale fuscous, somewhat paler ventrally, with yellowish tone. Anal brush rounded, fuscous, ventrally with longer, curved ivory tufts. Female abdomen fuscous, ventral side only slightly paler; ovipositor protruding.

Diagnosis. *S.palustris* (Z.) is larger, has more spread anal brush and a greater gloss; *tributella* (Z.) is usually smaller, more glossy, has curved labial palpi and is much paler on the dorsal side of the abdomen both in males and females; *picaepennis* (Hw.) and *disparella* (Tgstr.) are much larger and darker. Other similar species are darker than *paullella*.

Male genitalia (Fig. 71). Valva slightly curved, terminal half with a wide, extra lobe; uncus furcate; gnathos a straight spine; aedeagus very slender, slightly curved, thickest in middle; sternum VIII basically triangular, very extended distally and with a terminal pair of lobes.

Female genitalia (Fig. 110). Lamella postvaginalis ring-shaped with an anterior, extended sclerotization.

Distribution. Not in Denmark or Fennoscandia. Found in a few localities in C Europe, northernmost in the south of GDR (Schütze, 1897).

Biology. The larva was described by Schütze (1897). It is 8 mm long, dark olive brown, with regular yellowish marbling dorsally and laterally; dorsal line yellow, lateral line light brown, only faintly indicated. Intersegmental membrane of thoracic segments sulphur yellow, broadening laterally, ventral side of same colour as are also the whole of the segments with abdominal legs; otherwise olive green ventrally. Head dark brown; prothoracic shield slightly paler; anal plate and warts indistinct.

The larva is found in May on *Polytrichum,* on which it produces silken galleries. The larva pupates in a denser part of the web. After some two weeks the adult emerges. The flight period lasts from the end of May to the end of June.

The habitat of *paullella* seems to be sunny, open places on bouldery slopes with *Sedum album* L., Poaceae, mosses, etc., growing on soil between the boulders (Schütze, 1897).

10. *Scythris clavella* (Zeller, 1855)
Figs. 35, 72, 111.

Butalis clavella Zeller, 1855: 236.
Scythris villari Agenjo, 1971: 7. **Syn.n.**

13-15 mm (δ), 10-14 mm ($\female$). Head, labial palpi, antennae, collar, tegulae, thorax and forewing greyish green without a yellowish reflection. Many examples with numerous pale scales in forewing. Specimens with a white streak in the fold and a posterior whitish blotch are recorded from the southernmost region of the European USSR (Jäckh, 1978b). Hindwing 0.9 times as broad as forewing, uniformly fuscous. Cilia of forewing and hindwing dark fuscous. Male abdomen dark fuscous; anal brush divided into three distinct tufts. Female abdomen conspicuously tapered, ovipositor protruding; infuscated dorsally, but posterior margins of segments paler; lateral surface brownish, posteriorly with progressively more white scales, segments VI and VII becoming white. Tergum VIII, sometimes also sternum VIII, covered with brown and ochreous scales.

Diagnosis. *S.clavella* is very similar to *seliniella* (Z.) but is somewhat darker and lacks the yellowish gloss. *S.seliniella* usually has no pale scales on the forewing. Usually slightly smaller than *subseliniella* (Hein.) but impossible to distinguish from this species by external characters. *S.fuscoaenea* (Hw.) has a warm greenish gloss with an orange tint, not a cold greyish olive green. *S.productella* (Z.) is larger and darker, hindwings narrower and almost without gloss.

Male genitalia (Fig. 72). Very like *seliniella* (Z.), the only reliable difference is the shorter and thicker aedeagus with a small process one-third from base; terminal part of aedeagus slightly S-shaped.

Female genitalia (Fig. 111). Lamella postvaginalis pear-shaped, weakly sclerotized; ductus bursae widening posteriorly, about same length as tergum VIII, thereby distinguished from both *seliniella* (Z.) and *subseliniella* (Hein.).

Distribution. Not in Denmark or Fennoscandia. C Europe to SW Russia and the Caucasus, Italy, France and Spain. Closest to our area in C Poland and GDR (Thuringia).

Biology. Nothing is known of the immature stages. Adults fly from early June to mid-July, but earlier (mid-May) is S Europe.

Note. *S.villari* Agenjo was described from Spanish material (Agenjo, 1971) and said to differ from *clavella* in the valva, aedeagus and sternum VIII (saccus, according to

Agenjo). Agenjo compared *villari* with three male specimens of *clavella* (from Sarepta – the type locality) which showed distinct white markings in the forewing. *S.villari* is unicolorous and this led Agenjo to the opinion that *villari* was not conspecific with *clavella*. Passerin dËntrèves (1979) examined the type of *villari* but was unable to decide on its status. Comparing the illustrations and text of the original description with material of *clavella* the author suggests that *villari* is a junior synonym. Unicolorous specimens from France (Hautes Alpes) exhibit a rather slender valva and narrow tergum VIII, resembling *villari* in genitalia. The shape of the aedeagus, used as a specific character by Agenjo, may vary a great deal depending on the preparation. The female genitalia, illustrated by Agenjo, are difficult to interpret, but they seem to agree well with those of *clavella*.

11. *Scythris seliniella* (Zeller, 1839)
Figs. 36, 73, 112.

Butalis seliniella Zeller, 1839: 193.

14-15 mm (♂), 8-13 mm (♀). Head, labial palpi, antennae, neck tuft, collar, tegulae, thorax and forewing greenish grey with yellowish gloss. Female occasionally, male rarely, with whitish spots in forewing. Hindwing 0.9-1.0 times as broad as forewing, brown with narrow scales. Cilia of both forewing and hindwing fuscous. Male abdomen unicolorous dark fuscous. Anal brush tripartite, fuscous. Female abdomen stout, short and tapering. Dorsal surface of same colour as forewing; ochreous brown laterally; segments V and VI white.

Diagnosis. Very similar to *clavella* (Z.) and *subseliniella* (Hein.) but slightly paler and with more yellowish gloss. Somewhat smaller and with a stouter abdomen than *amphonycella* (Geyer) (not included in this work) which also has a compact anal brush. *S.fallacella* (Schl.) has a very stout abdomen and a much glossier forewing. A safe determination can only be achieved by examination of the genitalia.

Male genitalia (Fig. 73). Valva foot-shaped; uncus a short spine; aedeagus slender and almost straight; tergum VIII very weak with warty margin, median part rounded, terminal narrowing, bifid. The genitalia very much resemble those of *clavella* (Z.) but the aedeagus is longer than the valva.

Female genitalia (Fig. 112). Lamella postvaginalis bulb-shaped; antrum and posterior part of ductus bursae weakly sclerotized and wrinkled; anterior part of ductus bursae distinctly sclerotized, long and slender.

Distribution. Not in Denmark or Fennoscandia. Recently found in Latvia (Šulcs, 1981). Reported from GDR (Kyffhäuser) (Rapp, 1936), Poland, SW Russia, Hungary, Austria, Czechoslovakia, Switzerland, Italy, France and Spain. Also in Macedonia and Asia Minor.

Biology. Larva undescribed. *Peucedanum oreoselinum* (L.) Moench. has been suggested as a hostplant (due to the habitat-choice of the adults) as has *Genista* (Schütze, 1931). Rapp (1936) reported that the larva had been reared from *Cerastium brachypetalum* Pers.. The larva is supposed to be fully grown in May. Adults fly from early May to the end of June in sandy places with *Peucedanum oreoselinum* (Zeller, 1855) or in dune areas with *Festuca ovina* L., *Koeleria glauca* (Schrad.) DC., *Artemisia campestris* L., etc. (Šulcs, 1981). Rapp (1936) recorded the capture of many specimens near *Artemisia campestris*.

12. *Scythris subseliniella* (Heinemann, 1877)
Figs. 37, 74, 113.

Butalis subseliniella Heinemann, 1877: 439.
Scythris strouhali Kasy, 1962: 169.

14-16 mm (δ), 11-13 mm (φ). Head, labial palpi, antennae, collar, tegulae, thorax and forewing greenish brown or grey with dull reflection. Forewing often with pale scales, especially in specimens from the Iberian peninsula. Hindwing fuscous, 0.9 times as broad as forewing, with dense brownish, rounded scales; cilia fuscous as in forewing. Male abdomen unicolorous dark fuscous; anal tuft divided into three relatively well-defined brushes, somewhat paler than abdomen. Female abdomen of about same colour as forewing; two penultimate segments whitish ventrally.

Diagnosis. Male externally very similar to *seliniella* (Z.) and *clavella* (Z.); in *sub-seliniella* the forewing is usually duller and more pointed at apex, the costa slightly bent before middle. *S.fuscoaenea* (Hw.) is more bronzy, has a compressed anal tuft and darker and narrower hindwings. *S.fallacella* (Schl.) has a glossier forewing, a much stouter abdomen and a differently shaped anal tuft. Female specimens of *sub-seliniella, seliniella* and *clavella* are impossible to distinguish by external characters. Owing to infraspecific variation and similarity to other species (in both sexes), examination of the genitalia is often necessary to identify the species.

Male genitalia (Fig. 74). Valva slender, distal part wider, densely setose; uncus rounded; gnathos short, hooked; aedeagus long and slender, slightly S-shaped, thicker in anterior half; sternum VIII gently tapered, at end with two small flaps, separated by a shallow indentation, lateral margins straight. The shape of the valva clearly distinguishes *subseliniella* from *seliniella* and *clavella*.

Female genitalia (Fig. 113). Lamella postvaginalis as in *clavella* and *seliniella,* pear-shaped; ductus bursae long, as in *seliniella* but as broad as in *clavella* and with transverse wrinkles.

Distribution. Not in Denmark or Fennoscandia. Recorded from GDR (Thuringia), Austria, France, Spain, Italy and Czechoslovakia. The type series is from Oedenburg on the border between Austria and Hungary. Also reported from the Caucasus to Hungary (Falkovitsh, 1981).

Biology. Immature stages unknown. Adults fly in May and June, but in southern Europe they may also be collected in April and July. *S.subseliniella* is usually found in mountain areas.

13. *Scythris sinensis* (Felder & Rogenhofer, 1875)
Figs. 38, 75, 114.

Butalis sinensis Felder & Rogenhofer, 1875: CXL/11.
Staintonia? apiciguttella Christoph, 1882: 42.
Scythris pyrrhopyga Filipjev, 1924: 42.
Eretmocera pentaxantha Meyrick, 1929b (Vol. 3): 543.
Scythris kibarae Matsumura, 1931: 1094.
Scythris mitakeana Matsumura, 1931: 1095.

12-14 mm. Head, labial palpi, antennae, collar, tegulae, thorax and forewing very dark brown, almost black, with particular lighting faintly bluish. Hindwing 0.9 times as broad as forewing, dark brown. Cilia of both forewing and hindwing fuscous, sometimes with a trace of a cilia line in the former.

The nominate *S.sinensis sinensis* (Feld. & Rog.) from Shanghai, China exhibits distinct yellow-orange markings in the forewing, a blotch in the fold at one-quarter and a smaller one in the apical area, a morph seldom found in Latvia (Šulcs, in litt.; Sattler, 1971). Abdomen of both sexes hardly varying; terminal three-quarters dorsally and ventrally yellow-orange.

Diagnosis. *S.sinensis* is easily recognizable by its blackish forewing and bright-coloured, yellow-orange abdomen.

Male genitalia (Fig. 75). Valva slender, tapering and bent; gnathos tapered with increasing sclerotization towards blunt apex; aedeagus uniformly thick in basal three-fifths, then suddenly angled and irregularly tapered to a point.

Female genitalia (Fig. 114). Lamella postvaginalis a semi-circular, sclerotized disc, dentate posteriorly; rim of ostium bursae U-shaped; antrum well defined; sternum VII with a medial, large, sclerotized ellipse.

Distribution. Not in Denmark or Fennoscandia. Found in Latvia (Šulcs, 1973) and Lithuania (Ivinskis, in litt.), where appearing to be increasing its range. Otherwise known from China, Siberia, Japan and South Korea (Park, 1977).

Biology. Šulcs (in litt.) describes the larva as follows: length 10 mm, fusiform, pale grey or pale brown with mottled markings; feeding on *Chenopodium album* L. in July, hiding between spun leaves. Adults fly from the end of May to mid-July in the Baltic countries. The biology of *sinensis* in E Asia is described by Sattler (1971) and is here quoted: 'Larva on *Chenopodium album* L. var. *centrorubrum* Makino. Eggs deposited on buds. Larvae in web among buds or leaves, feeding on leaves, in autumn also on seeds. First and second instar larvae several in common web, third and fourth instar solitary. Pupa in dense cocoon. Hibernation in pupal stage. Three generations a year. First generation of moths (unmarked specimens) in May-June, second (marked specimens) in July, third (marked specimens) in September-October. There is no evidence for at third generation outside of Japan.'

14. Scythris productella (Zeller, 1839)*
Figs. 18, 39, 76, 115.

Oecophora productella Zeller, 1839: 193.

16-19 mm. Head, labial palpi, antennae, collar, tegulae, thorax and forewing dark brown with olive-green tinge. All scales in forewing of same shape and colour. Cilia fuscous. Hindwing 0.7 (male) times as broad as forewing or narrower (female), dark brown; average length of scales about four times their breadth, unicolorous. Cilia as in forewing. Male abdomen almost unicolorous, fuscous. Anal brush somewhat paler. Female abdomen longish and of almost even width, last segment somewhat tapered. Upper surface with fuscous and cream scales, posterior margins of segments with lighter scales. Anterior half of last segment with only ochreous-yellow scales, posterior half dark brown. The extent of the pale scales may vary. Ventral surface fuscous, but 2-3 last segments ochreous-yellow, except apex of last segment. Papillae anales often protruding.

Diagnosis. *S.productella* is smaller, darker and greener than *obscurella* (Scop.), the hindwing being considerably narrower than the forewing. *S.grandipennis* (Hw.) is less green, often with whitish scales on the forewing; the anal brush is curly. *S.fuscoaenea* (Hw.) is usually smaller and paler, with a glossier forewing; above eye with whitish scales, extending to scape. Unicolorous forms of *cuspidella* (Den. & Schiff.) usually have a red gloss in the basal area and a broader hindwing. In C and S Europe several similar species occur (*S.amphonvcella* (Geyer), *podoliensis* (Rbl.), *cupreella* (Stgr.),

Male genitalia (Fig. 76). Valva slender, tapered and irregularly curved; subapically with ventral group of bristles. Uncus horseshoe-shaped, inwardly with noticeably weaker sclerotization. Gnathos rather short, hooked. Aedeagus of moderate size, thick and straight in basal half, deflected at three-quarters, apex narrow and obliquely truncated. Sternum VIII with pair of spearhead-like sclerotizations.

Female genitalia (Figs. 18, 115). Lamella postvaginalis large, rounded or trapezoid: antrum short, weakly sclerotized; lamella antevaginalis spatular; sternum VII rounded and more strongly sclerotized posteriorly.

Distribution. In Finland found in the district of Ka. In Sweden in few northern places on the Gulf of Bothnia. Not in Denmark or Norway. Otherwise in C Europe, especially in mountain areas.

Biology. The larva is recorded as living on *Origanum vulgare* L. in spring (Schütze, 1931, and others). No description of the larva has been published. Swedish records imply a different hostplant as *Origanum* does not grow in the known finding places. Adults appear at the end of June to mid-July in stony or gravellary, sunny places, sparcely covered with *Hippophaë, Salix* spp., *Poaceae,* mosses and lichens and other trivial plants confined to dry, coastal places.

Note. The taxon *Butalis psychella* Zeller, 1839 is perhaps conspecific with *productella,* but its identity is still obscure. It seems to have been last mentioned in the literature by Zeller (1855).

15. *Scythris palustris* (Zeller, 1855)*
Figs. 40, 77, 116.

Butalis palustris Zeller, 1855: 217.
Butalis mattiacella Rössler, 1866: 355; Hannemann, 1958b: 84.

11-13 mm. Head, labial palpi, antennae, collar, tegulae, thorax and forewing greyish bronze, scales unicolorous, glossy. Labial palpus rather short, slightly erect. Hindwing 0.7 times as broad as forewing, fuscous. Cilia of forewing and hindwing brownish, infuscated, in hindwing with faint trace of cilia line. In his description of *mattiacella*, Rössler (1866) stated that the forewing shows scattered, long grey-white scales; such scales have not been observed in the Danish specimens examined, nor did Hannemann (1958b) refer to grey-white scales. Male abdomen brownish, infuscated and dull. Anal brush somewhat paler, brownish ochreous. Female abdomen conspicuously conical; coloured dorsally almost as in forewing, but somewhat darker, last segment yellowish; ventral surface of segment III-V whitish in middle, last segment very pale – cream; papillae anales protruding.

Diagnosis. *S.palustris* may easily be confused with some other concolorous *Scythris*. *S.tributella* (Z.) is smaller and darker, the hind tibia clearly paler inside; pecten bristles few or absent, anal tuft entire; *S.picaepennis* is darker and has a much broader hindwing; *S.laminella* (Den. & Schiff.) is much darker and fresh examples have a bluish tinge; *S.crassiuscula* (HS.) has a very stout abdomen; *S.paullella* (HS.) is paler and has a broader hindwing, the labial palpi shorter, almost straight, the anal tuft rounded.

Male genitalia (Fig. 77). Valva long, slender, bent in middle, apex drawn to a point. Uncus a curved, sclerotized band. Gnathos a warty tooth. Aedeagus short, distally tapering. Tergum VIII of characteristic shape, bifid posteriorly, base with boomerang-shaped process at each side. Sternum VIII broadly triangular, with curved strengthening-border, incurved at base.

Female genitalia (Fig. 116). Lamella postvaginalis a weak sclerotization, two weak flaps, furnished with bristles posteriorly; sternum VII with deep central slit, anteriorly with fine conspicuous bristles; posteriorly with a short unilateral process.

Distribution. *S.palustris* (Z.) has been found in two localities in Denmark, eastern Jutland (Hald, in 1916) and Fyn (Fåborg, in 1916 and 1922) (Larsen, 1916, 1927). In Finland *palustris* has occurred in some southern provinces. Not in Norway or Sweden. Otherwise reported from Austria (Hartig, 1964), Lithuania (Ivinskis, in litt.), Poland and Germany (Hannemann, 1958b), Switzerland (Sauter, in litt.), France (Leraut, 1980) and Hungary (Gozmány, in litt.). Maybe also in Czechoslovakia. The habitat is damp sites.

Biology. The egg has been superficially described by Schütze (1904): cream, long, sometimes irregular, with faint gloss and granular surface. The colour changes to a richer yellow and then reddish immediately before hatching. The larva, also according to Schütze (1904), is first yellowish, after some days becoming reddish with a blackish head, greyish prothoracic shield and anal plate. The mature larva is about 9 mm long, fuscous; dorsal side mottled with pale red-brown; laterally with a faint reddish tone; warts black; thoracic legs black, prolegs laterally with black, semicircular line; dorsal line pale, fine; subdorsal line reddish brown to blackish brown.

The larva feeds on moss *(Rhytidiadelphus squarrosus* L.) (Schütze, 1904) on which it makes white silken tubes. Pupation takes place in a web close to the moss stem. After two weeks the adult emerges. The pupa is orange with red around the segment articulations. Adults fly in June and July in moist places. The species apparently is very local and the female is especially difficult to obtain.

16. *Scythris muelleri* (Mann, 1871)
Figs. 41, 78, 117.

Butalis mülleri Mann, 1871: 81.
Scythris tolli Rebel, 1938: 105; Hannemann, 1960: 86.
Scythris mülleri var. *unicolorata* Osthelder, 1951: 183; Hannemann, 1960: 86.

9-10 mm. Head, labial palpi, antennae, collar, tegulae and thorax dark brown. Forewing blackish brown with faint reddish tint, faintly glossy; in fold and on cell end with small whitish or cream spots, the outer one rounded. The pale spots are occasionally missing (var. *tolli* Rbl.; cf. Hannemann, 1960). Hindwing 0.7-0.8 times as broad as forewing, dark brown, paler towards base. Cilia of forewing and hindwing dark fuscous. Male abdomen slender, widening posteriorly, ending in a very large and spreadout anal brush; this and the whole abdomen a very dark fuscous. Female abdomen cylindrical, only last segment tapering, dark fuscous dorsally, somewhat paler ventrally; posterior segment with medial depression ventrally.

Diagnosis. Easily distinguished from all other European species by the forewing markings, the huge male anal brush and the relatively small size. *S.punctivittella* (Costa) from S Europe and Asia Minor might possibly be confused with *muelleri* but the former is larger and the inner marking is in the form of a basal streak, though not exactly reaching the base.

Male genitalia (Fig. 78). Valva slender, broadening subapically; uncus furcate with long, slender prongs in V-arrangement, with lateral flaps near end; gnathos thick basally, abruptly tapered to obliquely truncated end; aedeagus short, terminally angled and with minute teeth; tegumen arched, posterodorsally with two extremely long, evenly broad processes; sternum VIII posteriorly furcate with a narrow neck; tergum VIII square, incurved posteriorly.

Female genitalia (Fig. 117). Lamella postvaginalis two curved sclerites, fused posteriorly and, sometimes, anteriorly; sternum VII with posterior hump; tergum VII indented, almost square; sternum VI short, longest medially.

Distribution. Not in Denmark or Fennoscandia. Recorded from Poland (Rebel, 1938), SW Russia, southern Germany (Osthelder, 1951), Czechoslovakia (Povolny, in litt.), Hungary and Austria (Glaser, 1962).

Biology. Immature stages unknown. Adults are found at the end of May and in

June. Glaser (1962) recorded finding this species in Burgenland, Austria. He described the habitat as a moist meadow with smaller, dry hillocks on which a number of moths were caught.

Note. Hannemann (1960) erected a new genus, *Parascythris*, with *muelleri* (Mann) as type-species. The genus is not recognized in the present work, pending a complete revision of the Scythrididae.

17. *Scythris inspersella* (Hübner, 1817)*
Figs. 8, 19-21, 42, 79, 118.

Tinea inspersella Hübner, 1817: 443.
Scythris hypotricha de Joannis, 1920a: 145.

14-17 mm. Head, labial palpi, antennae, collar, tegulae and thorax blackish. Forewing with sparsely scattered whitish scales, especially in outer half. One or two smaller groups of white scales in fold, sometimes forming a fine streak. Hindwing 0.8 times as broad as forewing, fuscous. Cilia of forewing and hindwing fuscous. Male abdomen blackish brown dorsally; ventral surface ivory to dull grey, anteriorly darkened by admixture of brown or grey scales. Dorsal side of female abdomen as in male; ventrally with pale grey to whitish scales except for the two terminal segments which are yellowish.

Diagnosis. *S.noricella* (Z.) is larger, paler and with a very long anal brush (male). *S.fuscopterella* Bengts. is smaller and with very few, not white but pale grey scales in a slightly paler forewing.

Male genitalia (Figs. 79). Interpreted and described by Sattler (1981). Valvae asymmetrical, each with a strong, lateral prong; uncus furcate, membranous, with setae; tegumen strongly asymmetrical, on the left with a short, blunt process, on the right with a long process which is apically widening and curved, at base fused with inside of tegumen; aedeagus long and slender, at base rather broad, gradually tapering, apically narrow, gently curved from middle; tergum VIII (not in fig. 79) crescentic, with warty, posterior margin; sternum VIII weak, warty, an inconspicuous strip.

Female genitalia (Fig. 118). Lamella postvaginalis weakly sclerotized, rhombic; lamella antevaginalis finger-shaped.

Distribution. In Denmark in all districts; in Sweden widespread, only absent from the alpine and subalpine regions; in Norway in four southern provinces; in Finland from most provinces north to the arctic circle. Widely distributed in Europe but not recorded from the Mediterranean area.

Biology. The larva was first described by Zeller (1855). It is dark reddish- or violet-brown with pale marbling; dorsal, subdorsal and subspiracular lines pale olive-green, indistinct; subventral line broad, green; belly pale olive-brown; head, prothoracic shield, anal plate and legs blackish; ecdysial line narrow white. Chaetation as shown in fig. 21. The larva feeds on *Chamaenerion angustifolium* (L.) and is found from mid-

May to mid-July; it is often gregarious. It entwines the upper leaves in a characteristic manner and, during growth, the top of the plant is bent sideways. It is incidently found together with larvae of *S.noricella* (Z.) (Svensson, pers. comm.). Pallesen & Palm (1975) report the finding of several larvae on *Potentilla;* this probably represents only a fortuitous stray occurrence. Pupation takes place in a white, transparent cocoon, fastened to a substrate near the ground, or sometimes in the spun, larval web. Occasionnally several cocoons may be spun together. Adults fly from early July to mid-August in all sorts of habitats where the foodplant grows. Exceptionally, the author has found the species being attracted to light.

18. *Scythris noricella* (Zeller, 1843)*
Figs. 16, 22, 43, 44, 80, 119.

Oecophora noricella Zeller, 1843: 151.

18-22 mm. Head, labial palpi, antennae, collar, tegulae and thorax fuscous. Labial palpi rather small. Forewing dull dark grey, somewhat brownish, with some apical admixture of whitish scales, especially in female. Scales slender, two- or three-pointed. In fold near base, in middle of forewing and on end of cell small blackish brown blotches. Cilia dark fuscous. Hindwing 0.8 times as broad as forewing, greyish or fuscous. Cilia fuscous with paler base. Specimens from Greenland are considerably smaller (14-16 mm) (Fig. 44). They are also paler, with more whitish scales and paler background (cf. Wolff, 1964). Male abdomen unicolorous fuscous with a paler anal tuft. Female abdomen greyish with a faint brown tinge, ventrally somewhat paler, almost silver-grey.

Diagnosis. Size, long anal brush and general habitus are specific and only *inspersella* (Hb.) may be confused with *noricella*. *S.inspersella* is, however, darker, almost black, and considerably smaller.

Male genitalia (Figs. 16, 80). Valvae very long, in basal half united and distally sickle-shaped, very pointed. Near base of valva two lip-shaped sclerotizations; aedeagus slender, curved and pointed; uncus bilobed, nodose; gnathos two flaps with minute projection i middle; tergum VIII long and narrow with parallel margins, scalloped terminally.

Female genitalia (Figs. 119). Apophyses posteriores very long; lamella antevaginalis a long plate with irregular folds and projections, anterior part broadly anchor-like with small wrinkles; sternum VII posteriorly concave, with rounded edges and straight anterior margin.

Distribution. In Sweden recorded from six northern provinces; in Norway recently found in Os; in Finland older records from five southern districts. Not in Denmark. Reported from mountain areas in eastern C Europe, Belgium and the Alps. Also in Greenland and Kamtchatka (Karsholt, pers. comm.).

Biology. Larva dorsally and ventrally pale olive-green, laterally dark olive-green. Dorsal line rather broad, indistinct, dull yellowish; subdorsal line same as dorsal line;

thoracic prolegs dark brown, inwardly yellowish; head dark brown, around ecdysial line pale yellow; prothoracic shield blackish brown, with distinct median division. Larva in June and July on *Chamaenerion angustifolium* (L.) (in Greenland *Chamaenerion latifolium* (de Lesse & Viette, 1949; Wolff, 1964)) in an extended web along the flower-stalk and the leaves, feeding on buds, flowers and leaves. Often several larvae live together and occasionally they are found together with the larva of *S.inspersella* (Hb.) (Svensson, pers. comm.). Adults occur from mid-July to mid-August in gravel-pits, along road-ways and similar places with high insolation. Unlike most other European *Scythris, noricella* is often attracted to light.

19. *Scythris empetrella* Karsholt & Nielsen, 1976*
   Figs. 9, 45, 81, 120.

Scythris empetrella Karsholt & Nielsen, 1976b: 151.
Glyphipteryx variella Stephens, 1834: 274, nom.praeocc.

8-10 mm. Head, antennae end neck tuft blackish, often with whitish scales and brown tinge. Collar, tegulae and, sometimes, thorax with greyish white scales, ground colour brownish black (fresh specimens). Forewing most often dark grey, occasionally with brown hue; white scales, usually forming a short streak near base, a median spot in fold, a tornal blotch sometimes extended towards a costal spot; basal and costal half of forewing normally without white scales in most specimens. Blackish marks often standing out between the three white spots. Exceptionally, white markings may be strongly reduced. Cilia dark greyish. Hindwing 0.7 times as broad as forewing, in male grey with very faint brown tint, in female considerably darker and browner. Cilia as in forewing but with pale brownish base. Dorsal side of male abdomen grey with admixture of scales which range over the whole spectrum of grey shades; ventrally whitish or pale grey, darkened anteriorly; anal tuft pale fuscous. Female abdomen dorsally pale grey with a faint bluish tone but first two segments black, rather glossy, and terminal segment dark brown; ventral surface as in male but last segment dark brown.

   Diagnosis. *S.empetrella* Karsh. & Niel. is larger than *siccella* (Z.) and *crypta* Hann. and has more extensive white markings in the forewing. *S.ericivorella* (Rag.) is dark brown without white scales and has a broader hindwing.

   Male genitalia (Figs. 81). Strongly reduced, valvae erect, diverging, gently narrowing and broadly rounded at apex; aedeagus small, angled close to obliquely truncated apex.

   Female genitalia (Fig. 120). Lamella postvaginalis asymmetrical, with two rounded posterior humps; left hump with small crescent-shaped lateral sclerotization; antrum weak, cup-shaped.

   Distribution. In Denmark found in some districts in Jutland and Zealand, also in Bornholm; from Sweden in the southern coastal areas and in Ång.; from Norway in Ø and VAy; from Finland in two southern provinces (Ab, N) and in the coastal part of Om and ObS. Otherwise in western C Europe; also in Great Britain.

Biology. Larva dull purple-brown with black head and brown prothoracic shield (Meyrick, 1928). According to the same author, and also to Hackman (1945) and Emmet (1979), the larva lives in long silken galleries attached to the roots, twigs and runners of *Empetrum* (in Great Britain and France also *Calluna* and *Erica)*. The silken tubes may often extend below the sand surface. The larva is found in April and May and apparently prefers bare sand. Adults appear over a long period, from the end of May to the end of July; they usually stay near the edge of the patch of foodplant, jumping or flying in the sunshine, and they are easily disturbed by gently sweeping one's hand over the hostplant. The moth closely resembles dead leaves of *Empetrum* and thus may be overlooked.

20. *Scythris siccella* (Zeller, 1839)*
Figs. 46, 82, 121.

Butalis siccella Zeller, 1839: 193.

8-10 mm. Head, labial palpi, antennae, collar, tegulae and thorax dark brown. Forewing dark brown, dull with some admixture of ochreous scales in apical half. Fold often with one or two small spots of white scales. Cilia brownish. Hindwing 0.6-0.7 times as broad as forewing, brown, darkened terminally. Cilia as in forewing. Male abdomen brownish grey dorsally, ventral side somewhat paler, dirty brown; anal tuft dusky ochreous. Dorsal side of female abdomen as in male but last segments paler with blackish brown posterior margins; underside whitish, anteriorly becoming darker ochreous.

Diagnosis. *S.crypta* Hann. has a more slender abdomen, almost blackish (not brownish) forewings and a narrower hindwing. *S.empetrella* Karsh. & Niel. generally has more white markings in the forewing. *S.ericivorella* (Rag.) has a broader hindwing, almost as broad as the forewing. Other similar species are considerably larger.

Male genitalia (Fig. 82). Strongly asymmetrical. Left valva short, truncate, with inturned projection; right valva slender, claviform; uncus and gnathos reduced to a broad plate with terminal, membranous hood; sternum VIII (not figured) bilobed, weak and slightly asymmetrical; aedeagus proportionally stout, tapered and curved at apex.

Female abdomen (Fig. 121). Lamella postvaginalis acute-angled with lateral, pointed projections; base with two symmetrical, sclerotized projections; with unmelanized 'windows'; sternum VII with prominent hump posteriorly, incised medially.

Distribution. From Denmark in Jutland and Bornholm; in Sweden old records from a few southern provinces but only specimens from Sk. have turned out to be correctly identified. Not in Norway or Finland (cf. Hackman, 1945). In Great Britain in the southernmost part (Meyrick, 1928). Records are also available from C and S Europe, but some of these might also be dubious.

Biology. Larva according to Spuler (1910) reddish white with brown head and prothoracic shield. Meyrick (1928) gave a more detailed description; larva slender, dull purple; incisions paler, between segment 2-6 whitish; spots sometimes white; head and prothoracic shield black-brown. Many foodplants have been suggested: *Armeria maritima* (Mill.) Willd. (Spuler, 1910); *Hieracium, Plantago, Helichrysum, Helianthemum, Ononis, Scabiosa columbaria* L., growing close to ant nests (Schütze, 1931). Petry found the larva in April under the basal leaf rosette of *Scabiosa columbaria* in webtubes on the ground (Rapp, 1936). Spuler (1910) also mentioned *Rumex acetosa* L. for *variella* Stph. but this observation should perhaps be referred to *siccella*. Emmet (1979) gives *Thymus drucei* Ronn., *Lotus corniculatus* L., *Plantago* spp. and *Cerastium* spp. as foodplants. The larva mines leaves from a long silken tube on or below the surface of the sand, attached to half-buried stems of the hostplant. Adults appear in open, sandy places with sparse, low vegetation from the end of May to mid-July.

21. *Scythris tributella* (Zeller, 1847)
Figs. 47, 83, 122.

Oecophora tributella Zeller, 1847: 833.
Oecophora terrenella Zeller, 1847: 834.
Oecophora parvella Herrich-Schäffer, 1855: 265.
Oecophora denigratella Herrich-Schäffer, 1855: 271.
Butalis serella Constant, 1885: 11; Passerin d'Entrèves, 1980: 52.
Scythris karnyella Rebel, 1918: 87.
Scythris monotinctella Turati, 1924: 180.
Scythris bulbosella Lhomme, 1949: 795.
Scythris igaloensis Amsel, 1951: 419.

9-11 mm. Head, labial palpi, antennae, neck tuft, collar, tegulae, thorax and forewing greyish olive, with moderate, yellowish or reddish gloss. Appeerence variable; sometimes paler scales in fold of forewing; specimens from Italy may be almost black. Hindwing 0.6-0.7 times as broad as forewing, with dense, oval scales, fuscous with faint violet gloss. Cilia of forewing and hindwing fuscous. Male abdomen slender, dorsally dark grey, ventrally paler, ochreous or grey, posterior segments even whitish in some specimens; anal tuft short, thin, compact, somewhat paler than abdomen. Female abdomen dark fuscous, slightly paler posteriorly; ventral surface greyish ochre anteriorly, two penultimate segments with white extended dorsolaterally; last segments completely fuscous.

Diagnosis. Similar to *S.laminella* (Den. & Schiff.) but with a broader hindwing and with inner side of hind tibia whitish. *S.crassiuscula* (HS.) is considerably darker and has a much stouter abdomen; *palustris* (Z.) has a more slender abdomen, more glossy forewing and paler fringes; *paullella* (HS.) has a broader forewing (1.3-1.4 mm against 1.1-1.2 mm) and is larger.

Male genitalia (Fig. 83). Valva cygnate; uncus, weak, warty; aedeagus comparatively short, broad, bent and pointed; tergum VIII with sclerotized, rugose area posteriorly; sternum VIII bifid posteriorly, with sclerotized arches and transverse ridge.

Female genitalia (Fig. 122). Very much resembling those of *crassiuscula* (HS.) (Fig.

128). Lamella postvaginalis triangular with posterior tip truncated; lateral sclerotized border absent or narrow. Sternum VII with posterior margin rounded, uneven; apophyses anteriores twice as long as lamella postvaginalis.

Distribution. Not in Denmark or Fennoscandia, nor in Great Britain. C and southern E Europe, closest to the Scandinavian area at Braunschweig i GDR. Prefers lush chalky slopes. *S.tributella* was reported from Scania, S Sweden, as *parvella* (HS.) by Wallengren (1875). Re-examination has shown, however, that the specimen in question is *laminella* (Den. & Schiff.) (Svensson, 1978).

Biology. Larva undescribed. Schütze (1931) stated that *tributella* (Z.) *(parvella* (HS.)) has been reared from *Coronilla varia* L.. There are also indications that other foodplants such as moss, grass or *Cerastium,* may be utilized. Adults fly from May to August, possibly in two broods in the south. In Italy *tributella* may be found as late as September (Jäckh, in litt.).

22. *Scythris picaepennis* (Haworth, 1828)*
Figs. 10, 48, 84-88, 123.

Porrectaria picaepennis Haworth, 1828: 536.
Oecophora senescens Stainton, 1850: 22; Bradley, 1966: 137.
?Oecophora vagabundella Herrich-Schäffer, 1855: 269.
Butalis aeneospersella Rössler, 1866: 353; Hannemann, 1958b: 82.
Scythris heterodisca Meyrick, 1929a: 149; Pierce & Metcalfe, 1935: 49.
Scythris joannisella Le Marchand, 1938:131.

10-12 mm. Head, labial palpi, collar, tegulae and thorax with dark fuscous scales with pale bases and dark purplish tips. Labial palpi and tips of tegulae often with completely light brown and grey scales. Upper margin of eye sometimes crowned with whitish hairs. Under high magnification the wing surface seems mottled because of the pale bases and dark blue-purplish tips of scales. Female, more seldom male, with light scales appearing in strong contrast to dark ground colour. Hindwing as broad as forewing, dark brown. Cilia of forewing and hindwing fuscous. Male abdomen dark fuscous with slightly paler scales on ventral surface. Female abdomen dark fuscous on dorsal side. Two penultimate segments whitish ventrally. Sternum V often more or less clothed with ivory scales.

Diagnosis. In Fennoscandia *picaepennis* (Hw.) may primarily be confused with *disparella* (Tgstr.) which never has white scales above the eye or pale scales in the forewing. *S.laminella* (Den. & Schiff.) is smaller and has narrower hindwings, only about half as broad as the forewing. *S.potentillella* (Z.) has a broader forewing with usually many white scales, or else is very similar and only distinguishable by genitalia; the female of *potentillella* sometimes has white scales in the fold, forming a streak which *picaepennis* lacks. *S.crassiuscula* (HS.) is more glossy and has no pale scales in the forewing.

Male genitalia (Figs. 84-88). Very compact and similar to those of *disparella* (Tgstr.)

but valva longer and somewhat narrower. Uncus variable, truncate to furcate (may be rendered visible by removing anal tuft with a stiff brush). Aedeagus short, apically slightly bent and blunt-ended.

Female genitalia (Figs. 123). Very similar to *disparella* (Tgstr.). Lamella postvaginalis triangular, truncate posteriorly and concave anteriorly. Segment VIII with a row of broad postmarginal scales (not drawn) (see Benander, 1951). Apophyses anteriores very short and hooked. Sternum VIII anteriorly broad, trapezoidal.

Distribution. In Denmark in most districts; in Sweden only reported from four southern provinces; recorded from VAy in Norway (Opheim, 1978). In Finland in the southern part north to about latitude 62°. Also reported from Great Britain, C Europe and N Africa. In the Alps *picaepennis* is found also at high altitude.

Biology. The larva and its biology are described by Benander (1965). It is bluish grey with light yellow dorsal and subdorsal lines; head yellow, brownish behind; prothoracic shield black with yellow front; legs dark green; warts black; spiracles red-circled. The larva is found in June and early July on *Lotus* in silken galleries amongst moss with webbing up the shoots. Adults appear from mid-June to the end of July.

In Great Britain the larva is also found on *Thymus* in May (Pierce & Metcalfe, 1935a). Schütze (1931) reports, along with *Lotus*, also *Thymus, Helianthemum, Succisa* and *Plantago* as foodplants.

Note. The taxon *vagabundella* (Herrich-Schäffer, 1855) has for a long time appeared in the literature as a good species. Herrich-Schäffer's type material has not been available and is possibly lost, but old specimens labelled *vagabundella,* examined by Jäckh (in litt.) and the author, have turned out to be *picaepennis,* and this identity was actually suggested already in Heinemann & Wocke (1877). Falkovitsh (1981) considered *vagabundella* a senior synonym of *flavidella* Preiss.; this view appears unwarranted.

23. *Scythris disparella* (Tengström, 1848)*
Figs. 49, 89, 124.

Oecophora disparella Tengström, 1848: 121.

12-14 mm. Head, labial palpi, antennae, collar, tegulae and thorax dark brown. Forewing of almost same colour, but especially at costa and at apex more purplish brown. Forewing with smooth appearance due to almost unicolorous, narrow and glossy scales. Cilia dark fuscous. Hindwing 0.9 times as broad as forewing, dark brown, very faintly purplish-tinged. Cilia as in forewing.

Diagnosis. *S.disparella* has a more glossy appearance than the closely related *picaepennis* (Hw.) which has rather rough-scaled forewings. The scales of *picaepennis* are purplish black with conspicuously paler bases. Frequently *picaepennis* has dirty yellowish scales which obtrusively contrast with the dark background. Such scales are absent in *disparella*. Often *picaepennis* has whitish scales above the eye, those are never present in *disparella*. Male abdomen unicolorous dark fuscous. Anal brush small, pointed, fuscous. Female abdomen as in male but two penultimate segments ivory on underside.

Male genitalia (Fig. 89). Very compact. Valva rounded and very short. Uncus regularly tapering, ending in a pointed or weakly rounded tip (may be rendered visible by brushing off the anal tuft) (cf. *picaepennis* (Hw.)). Aedeagus short, tapered and pointed.

Female genitalia (Fig. 124). Very similar to *picaepennis*. Lamella postvaginalis smaller, not as pronouncedly triangular; apophyses anteriores straight; postero-lateral flaps on tergum VIII large, rounded, with distinct but minute warts; anterior trapezoid process of tergum VIII smaller and narrower than in *picaepennis* (cf. Benander, 1951).

Distribution. Recorded from the southern parts of Finland and from Sweden north to Ång. In Norway only in HEs (Aarvik, in litt.). Not in Denmark or Great Britain. Also found in Germany, the Netherlands, Belgium, France, Yugoslavia and Switzerland.

Biology. Immature stages unknown. Adults fly at the end of May and in June on sunny meadows, glades and slopes which are covered with low vegetation of *Lotus corniculatus* L., *Hieracium pilosella* L., *Fragaria vesca* L., *Rumex acetosella* L., etc..

24. *Scythris fuscopterella* Bengtsson, 1977*
Figs. 50, 90, 125.

Scythris fuscopterella Bengtsson, 1977: 55.

13-14 mm. Head, labial palpi, antennae, collar, tegulae, thorax and forewing dark fuscous, scales with blackish tip and paler base. Fold and apical area often with grey or whitish scales, especially in female. Near end of fold and above tornus small black spots. Hindwing 0.7 times as broad as forewing, fuscous. Cilia of forewing and hindwing infuscated brown. Male abdomen dorsally concolorous with thorax, very dark grey ventrally; anal tuft greyish ochre. Female abdomen dorsally as in male, ventral side greyish ochre, of richer colour towards hind margin of each segment; anal tuft very dark brown to blackish.

Diagnosis. *S.inspersella* (Hb.) is larger, darker, almost black, with more whitish, scattered scales. *S.noricella* (Z.) is even larger. Other similar species are smaller or have glossy forewings.

Male genitalia (Fig. 90). Valvae asymmetrical, abruptly narrowing, terminal parts almost of uniform breadth and with rounded tips. What might be uncus is a furcate structure with stout, hooked tips. In central part of genital armature a square plate, slightly emarginated posteriorly; plate, tips of valvae and uncus very strongly sclerotized. Sternum VIII two small humps on a sclerotized bridge between valvae. Aedeagus thin, curved in terminal half and pointed.

Female genitalia (Fig. 125). Lamella postvaginalis drop-shaped, with median 'window' and sublateral, symmetrical cavities. Tergum VIII in ventral view almost square.

Distribution. From Sweden in Ång., Vb., Nb. and T.Lpm; from Finland in Ks and

LkE. Not recorded elsewhere (see Note below).

Biology. Immature stages unknown. The common features of this species' habitats are that they are well-drained, sandy or gravelly places with rather sparse vegetation. The foodplant probably belongs to the Ericaceae. Adults appear from the end of June to the end of July and are very difficult to obtain.

Note. Jäckh (unpubl.) has examined a male specimen from France (Hautes Alpes, Refuge Tuckett, Glacier Blanc, 18.VII.1947, leg.coll. Buvat, Marseille (prep. Buvat 9579)), the genitalia of which closely resemble those of *fuscopterella*. However, there are small differences: aedeagus longer, uncus with longer projections, armature broader. Only further collecting can reveal if this French specimen is conspecific with *fuscopterella*, and if so, whether it represents a distinct geographical race or merely is an individual variant.

25. *Scythris braschiella* (Hofmann, 1897)
Figs. 51, 91, 126.

Butalis braschiella Hofmann, 1897: 241.

7-10 mm. Face with light brown scales, crown dark brown. Labial palpi ochreous, rather short, directed forward. Antennae conspicuously short, about half as long as forewing, dark brown. Collar, thorax and tegulae dark brown, the latter with pale anterior scales. Forewing dull, coarse-scaled, especially apically. Ground colour dark brown to blackish grey, fold with blackish streak splitting in middle of wing into one streak towards tornus and one towards apex. Above tornus a blotch of pale ochreous brown scales, sometimes anteriorly extending along the posterior margin. Scales in apical area black-tipped. Cilia brownish, somewhat infuscated. Hindwing 0.7 times as broad as forewing, brownish, with egg-shaped scales. Cilia as in forewing. Male abdomen dorsally fuscous, laterally with a few whitish scales; ventral side paler, ochre, darker anteriorly; anal tuft almost absent. Female abdomen as in male but dorsolaterally often with more whitish scales; ventrally with ivory scales in anterior halves of penultimate segments.

Diagnosis. Differing from all known European species in the rough-scaled forewing with ochreous brown blotches. Worn specimens may be similar to *siccella* (Z.) or *empetrella* Karsh. & Niel. but the stout abdomen almost without an anal tuft and the lack of white scales in the forewing distinguish *brachiella* from the former species.

Male genitalia (Fig. 91). Valvae slightly asymmetrical, long; uncus weak, bristled; gnathos a semicircular disc; aedeagus small, bottle-shaped; tergum VIII M-shaped with transverse bar; sternum VIII bow-shaped. The morphology of the male genitalia is somewhat reminiscent of that of *laminella* (Den. & Schiff.).

Female genitalia (Fig. 126). Also similar to those of *laminella*. Lamella postvaginalis bell-shaped, with comparatively large ventral slit; apophyses anteriores and posteriores of subequal length.

Distribution. Not in Denmark or Fennoscandia. Only known from N Germany (Mecklenburg and Brandenburg).

Biology. The only information from the literature is that in the original description. The immature larva is orange with a yellowish head, prolegs and anal plate; prothoracic shield faintly brownish. Later the head and anal plate become pale brown; prothoracic shield blackish brown with broad, median division. The mature larva is reddish yellow with indistinct, fine whitish dorsal line and fine, somewhat darker stigmata; head and prothoracic shield black-brown, the latter with a white, narrow, anterior seam; prolegs and rounded anal plate brown; bristles long, whitish, few; warts of same colour as body.

The larva feeds in July on *Armeria maritima* (Mill.) Willd. but is also found in June, probably after hibernation. The phenology of the larva is unclear and either there are two broods or only one prolonged brood. The larva makes a silken web which partly covers the basal leaves, eats these and, later, also the stem. During the last instars the larva produces denser galleries. Pupation takes place in a dense, oval, silken tube, fastened to a leaf of the hostplant or to moss or grass, and covered with sand grains. The pupa is pale brown.

Adults appear mainly from the end of July to early September and, at least in favourable years, also in June. They seem to be indisposed to fly; rather they jump over sandy ground and may be collected only on hot days. Unlike most other *Scythris* species, the females seem to be considerably more abundant than the males.

26. *Scythris laminella* (Denis & Schiffermüller, 1775)*
Figs. 2, 3, 11, 13-15, 52, 92, 127.

Tinea laminella Denis & Schiffermüller, 1775: 140.
Butalis succisae Rössler, 1866: 353; Hannemann, 1958b: 85.
Butalis schuetzei Fuchs, 1901: 383; Hannemann, 1958a: 65.
Scythris disqueella Fuchs, 1903: 62 (partim); Hannemann, 1958a: 65.

9-11 mm. Head, labial palpi, antennae, collar, tegulae, thorax and forewing very dark brown, with moderate, purplish gloss, especially near apex. Hindwing dark fuscous, strikingly narrow, being only 0.5-0.6 times as broad as forewing. Cilia of forewing and hindwing dark fuscous. Male abdomen slender, unicolorous blackish brown. Anal tuft infuscated, small, trifid. Female abdomen as i male but two penultimate segments dirty yellowish ventrally with admixture of brown scales anteriorly. Sometimes with pale anterior scales dorsally on terminal segment.

Diagnosis. *S.laminella* is recognizable amongst the Danish and Fennoscandian species by its size, slightly ascending labial palpi, narrow hindwing and gloss on the forewing. *S.picaepannis* (Hw.) and *potentillella* (Z.) are larger and normally have pale scales in the forewing; *S.siccella* (Z.) and *crypta* Hann. are smaller and lack forewing gloss; *S.ericivorella* (Rag.) has considerably broader hindwings and also lacks forewing gloss; *S.crassiuscula* (HS.) has strong forewing gloss and a thicker abdomen;

S.tributella (Z.) and *palustris* (Z.) are paler, have a yellowish gloss in the forewing, the former has broader hindwing.

Male genitalia (Fig. 92). The slender shape of the valva can be seen by brushing off the anal tuft. Valva of uniform breadth from a broad and angulate base, apex abruptly tapered, outer surface of terminal part of valva partly covered with broad, dark brown scales; uncus a warty, lip-shaped sclerotization, aedeagus sickle-shaped; tergum VIII (not drawn) triangular; sternum VIII (not drawn) quadrangular, with small emargination posteriorly.

Female genitalia (Fig. 127). Similar to those of *braschiella* (Hofm.); lamella postvaginalis broader, extensively unsclerotized medially, but with a minute posteromedian sclerotization. Apophyses posteriores considerably longer than apophyses anteriores.

Distribution. In Denmark found in NEZ; from many provinces in S Sweden, along the east coast north to Vb.; in two provinces in SE Norway; in southern half of Finland. Otherwise in E, C and W Europe (except Great Britain) southwards to N Italy.

Biology. Egg first very pale greenish yellow, later pale orange; surface with network of undulating ridges. In captivity eggs were placed on the basal glands of *Hieracium pilosella* L.. According to Spuler (1910) the larva is reddish brown with a black head and prothoracic shield (described as *schuetzei* (Fuchs)). Fuchs (1901) gave a more detailed description (here somewhat modified) of the larva of *schuetzei:* 9 mm long, fusiform, dull; head, prothoracic shield and anal plate black, faintly glossy; anal plate with large anterior and posterior projection and smaller, lateral projections; mandibles paler. The young larva is dark olive-brown, later paler olive-brown, dorsally with darker marbling. Dorsal line faint, pale; subdorsal line red-brown. Legs black, pale-ringed at articulations. According to Fuchs (loc.cit.) the larvae feed on *Rhytidiadelphus squarrosus* L., living in a silken web between mosses at the beginning of June. Schütze (1931) stated that the larvae feed on *Hieracium pilosella* L. under a thin web on the upper side of the leaves in April and May. No biological observations of the species in Scandinavia have been recorded.

The pupa is short and thick, reddish yellow, with three dark dorsal lines; in a fine, white web.

Adults fly from early June to the end of July on open, dry heaths where *Hieracium pilosella* grows.

27. *Scythris crassiuscula* (Herrich-Schäffer, 1855)
Figs. 53, 93, 128.
Oecophora crassiuscula Herrich-Schäffer, 1855: 268.
Scythris fuscocuprea auct., nec. Haworth (1828).
Scythris disqueella Fuchs, 1903: 62 (partim); Hannemann, 1958a: 65.
Scythris fletcherella Meyrick, 1928: 724.

9-12 mm. Head, antennae, collar, tegulae, thorax and forewing dark olive-brown with a yellowish or orange gloss, forewing with moderate metallic gloss. Labial palpi ascending, somewhat paler on upper surface. Hindwing 0.8 times as broad as forewing, with dark fuscous, oval scales; darker terminally. Cilia of both wings fuscous. Colours of female usually slightly darker than in male. Male abdomen dark brown with greyish tinge, somewhat glossy, ventrally with paler scales. Anal tuft indistinctly trifid, short and greyish. Female abdomen dorsally blackish brown, ventrally greyish ochreous, penultimate segment dirty ivory.

Diagnosis. *S.crassiuscula* (HS.) is distinguished from *potentillella* (Z.) and *picaepennis* (Hw.) by its stronger gloss and absence of slender, pale scales in the forewing; it is distinguished from *tributella* (Z.) and *laminella* (Den. & Schiff.) by its broader hindwing and from *disparella* (Tgstr.) by its smaller size and yellower gloss in the forewing. *S.fallacella* (Schl.) is considerably larger.

Male genitalia (Fig. 93). Uncus rounded; gnathos predominant, stout, with thornlike process at end. Aedeagus minute, bent, swollen in middle. Valva diminutive, weak sclerotized. Tergum VIII bifid with slightly incurved, rounded branches; base widening and with granular blotch in middle. Sternum VIII with deep and rounded cleft, basally with sclerotized bars.

Female genitalia (Fig. 128). Somewhat reminiscent of the genitalia of *S.tributella* (Z.). Lamella postvaginalis triangular with truncated apex, lateral border heavily sclerotized.

Distribution. Not in Denmark or Fennoscandia. Knowledge of the distribution of this species is probably incomplete owing to earlier misidentifications. For a long time the species has been known from southern England, and from C Europe with its main centre in the Alps. Closest to the Fennoscandian area in C Germany (at Mannheim).

Biology. Larva dull greenish brown with cream subspiracular line. Head and prothoracic shield dull yellowish, laterally marked with blackish (Meyrick, 1928). It is found on *Helianthemum* in May and June in a delicate web on the flowers (Schütze, 1931) or among stems (Meyrick, 1928). Adults fly from June to early August.

Note. *O.crassiuscula* was described from one ♂, one ♀, West Germany, Mombach near Frankfurt/Main, 19.VI. *(Schmid)*. The two syntypes could not be traced in the British Museum (Natural History), London, or in the Zoologisches Museum der Humboldt-Universität, Berlin. The identity of *crassiuscula* is here based on the first revisor, Zeller (1855: 224), who examined Herrich-Schäffer's male and stated it to be conspecific with *'fuscocuprea'* (=*fletcherella* Meyrick).

Porrectaria fuscocuprea Haworth, 1828, is currently considered to be a junior subjective synonym of *Monochroa tenebrella* (Hübner, (1817)) (Gelechiidae); this synonymy was confirmed by Bradley (1966: 137), who examined an original Haworth specimen and designated it as the lectotype.

28. *Scythris ericivorella* (Ragonot, 1881)*
Figs. 54, 94, 129.

Butalis ericivorella Ragonot, 1881: CXX.

9-11 mm. Head, labial palpi, antennae, collar, tegulae, thorax and forewing dark brown, infuscated and very faintly glossy. Hindwing 0.9 times as broad as forewing, fuscous, paler towards base. Female hindwing darker with richer brown colour. Cilia of forewing and hindwing fuscous. Male abdomen dorsally dark fuscous, ventrally somewhat paler. Scales covering valva dirty beige. Female abdomen blackish brown, considerably darker than forewing. Ventral surface paler. Ovipositor projecting.

Diagnosis. *S.laminella* (Den. & Schiff.) has a more glossy forewing, apically blue-tinged, and an almost unicolorous and narrower hindwing. *S.siccella* (Z.) and *crypta* Hann. are smaller and have a considerably narrower hindwing. *S.crassiuscula* (HS.) and *tributella* (Z.) have glossier forewings.

Male genitalia (Fig. 94). Valva minute, a small fold with bristles; uncus a sclerotized plate, posteriorly furcate, with additional flaps; gnathos well-developed, key-shaped, with open basal loop, middle region funnel-shaped, wrinkled, tip truncate with stout thorn-like process; aedeagus a truncated hook; sternum VIII with deep incision, forming two large, rounded projections; tergum VIII proportionally small, terminal pointed projection keel-shaped.

Female genitalia (Fig. 129). Lamella postvaginalis triangular, concave anteriorly and with subapical groove, and with a double pair of sclerotizations in anterolateral position.

Distribution. In Denmark found in western and northern parts of Jutland (Pallesen & Palm, 1974). Not in Scandinavia or Finland. Recorded from N Germany (at Rends-burg), Holland and France. On moors near the coast.

Biology. Ragonot (1881) found small larvae in May on *Erica cinerea* L., living in withered flowers. He gave no description of the larva. Spuler (1910) stated that the larva was reddish brown with a black head and prothoracic shield, living on *Erica tetralix* L. in May. Adults emerge from mid-June to the end of July; the habitat is flat areas with *Erica,* often immediately behind the dune zone.

Note. The closely related *S.insulella* (Staudinger, 1859) (not treated here) has very similar genitalia but has a broad whitish basal streak in the forewing. It is found in SW Europe.

29. *Scythris crypta* Hannemann, 1961*
Figs. 55, 56, 95, 130.

Scythris crypta Hannemann, 1961: 308.

8-9 mm. Head, antennae, collar and tegulae dark brown. Labial palpus with white basal segment, second segment pale in basal half, otherwise dark brown; terminal segment dark brown, dorsally somewhat paler. Thorax and forewing very dark brown, almost black, with faint bluish or purplish gloss. Terminal part of forewing rough-scaled; scales with considerably paler bases but no scales are entirely pale; whitish scales usually forming a spot in centre of wing. Cilia blackish brown. Hindwing 0.7 times as broad as forewing; hind margin almost straight; scales brownish, denser at apex. Cilia as in forewing, blackish brown. Male abdomen blackish brown with purplish tinge, ventrally greyish ochreous with scattered brown scales; anal tuft small, fuscous, compact. Female abdomen dorsally brownish black, ventral surface ivory, darker posteriorly, last segment dark brown; ovipositor usually protruding.

Diagnosis. *S.crypta* is very similar to *siccella* (Z.). No reliable external feature has been found to separate the two, but generally *crypta* has darker forewings, without paler scales in the apical area; the hindwing in *crypta* has a less curved (or even straight) hind margin; the dorsal surface of the labial palpus of *crypta* has darker scales than in *siccella* and the abdomen slightly slenderer. In the abdomen of the *crypta* female the contrast between the dark dorsal surface and the pale ventral surface is smaller than in *siccella*. *S.ericivorella* (Rag.) has no white scales in the forewing; it is larger and has a considerably broader hindwing, almost as broad as the forewing.

Male genitalia (Fig. 95). Valvae triangular, short, strongly sclerotized, fused basally; uncus indistinct; gnathos strongly sclerotized, rounded; aedeagus small, expanded basally, swollen medially, narrowed at apex; tergum VIII bifurcate, processes posteriorly more sclerotized towards apices; sternum VIII bilobed.

Female genitalia (Fig. 130). Lamella postvaginalis weak with a posterior thin arch and a small tongue in asymmetrical position; sternum VII with sclerotized, truncate posterior process.

Distribution. First discovered in SW Yugoslavia (Macedonia) at high altitude (Hannemann, 1961); then found in SW Sweden (Hall.) in 1978 (Svensson, 1979). Also reported from Italy (Passerin d'Entrèves, 1980). Not recorded from elsewhere.

Biology. Immature stages unknown. In Sweden, *crypta* is found on heathland with, among other plants, *Genista pilosa* L. and *Calluna vulgaris* (L.) Hull.. The former plant is probably the hostplant as silken webs, probably produced by the larva of *crypta,* have been found under the plant. Adults have been collected at the end of May to mid-June on carpets of *Genista*. There is possibly a second brood in August, but this has not been confirmed. Apparently *crypta* prefers burnt-off places where *Genista* initially meets less competition from other plants, chiefly *Calluna*.

30. *Scythris restigerella* (Zeller, 1839)
Figs. 57, 96, 131.

Butalis restigerella Zeller, 1839: 193.

(14)-16-18 mm. Head brownish with white scales above eye. Labial palpus fuscous, inner surface with white scales. Antennae, collar, tegulae and thorax fuscous. Forewing greyish with olive-brown tinge. Fold with numerous white scales forming a gradually broadening streak which ends at tornus in a diffuse patch of scattered white and beige scales. Cilia dark fuscous, at apex with white scales anteriorly. Hindwing 0.8 times as broad as forewing, dark brown, slightly lighter towards base. Cilia dark fuscous. Male abdomen dark fuscous dorsally, ventrally dirty ivory with variable admixture of somewhat darker scales. Female abdomen dull fuscous on dorsal surface, laterally with pale ochreous scales, anteriorly almost ivory, posteriorly brown; on ventral surface blotches of pale brown scales in pairs on a pale-ochreous ground.

Diagnosis. *S.restigerella* may be confused with *dissimilella* (HS.) and *vartianae* Kasy (the latter species not dealt with here) but the ground colour of the forewing is darker and less grey; the median streak in *restigerella* is more distinct and no whitish scales appear outside the streak. Certain specimens of the S European species *S.salviella* Jäckh and *scipionella* (Stgr.) may also be similar to *restigerella* but usually both species exhibit a distinct blotch of white scales on the tornus and a small, pale costal spot at three-quarters from base; in *restigerella* the white median streak runs into the fold, in *salviella* and *scipionella* it runs above the fold.

Male genitalia (Fig. 96). Genitalia reduced and degenerate as in *dissimilella* (HS.). *S.restigerella* has a comparatively shorter gnathos with only three terminal spines; the aedeagus has a basal bulge, posteriorly the bulge is abruptly S-shaped; in terminal two-thirds only slightly curved; sternum VIII has very broad plates, with posterior margins rounded, not straight as in *dissimilella;* tergum VIII is shallowly emarginate for less than half of its length.

Female genitalia (Fig. 131). Ductus bursae long, well sclerotized, comparatively regular with smooth surface, thereby differing from that of *dissimilella* (HS.) (Fig. 132).

Distribution. Not in Denmark or Fennoscandia. C and E Europe, often in mountain areas. Closest to Scandinavia in Poland (Romaniszyn *in* Schille, 1930).

Biology. The larva was described by Krone (1905): about 16 mm long, wax yellow, sometimes brownish yellow; dorsal, subdorsal and lateral lines of yellowish or brownish dots; head small, yellowish or brownish; prothoracic shield somewhat paler than head, both with black dots. The larva feeds in June and July on *Helianthemum* on which silken galleries are made. The pupa is yellowish brown and is in a web to be on the ground. Adults appear from mid-June to early September; also from mid-May to mid-June (Zeller, 1855) which suggests two broods, as given by Heinemann (1877) and Spuler (1910).

Note. *Oecophora vittella* Costa (1836) has sometimes been regarded as a senior synonym of *restigerella* (Hartig 1939, 1964). Neither of the two syntypes of *vitella* has an abdomen (Passerin d'Entrèves, in litt.) and it remains an unsettled problem whether this taxon is identical with *restigerella* or with *S.leucogaster* (Mann, 1872).

31. *Scythris dissimilella* (Herrich-Schäffer, 1855)
Figs. 58, 97, 132.

Oecophora dissimilella Herrich-Schäffer, 1855: 265.

14-16 mm. Head, labial palpi, antennae, collar, tegulae, thorax and forewing greyish brown, often with the addition of many white or whitish scales. Forewing with white scales concentrated around fold and in tornal area, in many specimens forming a fuzzy streak with some dark spots at hind margin of streak; remainder of forewing with many scattered whitish scales; females often with more whitish scales in forewing. Cilia infuscated. Hindwing 0.9 times as broad as forewing, brownish, considerably darker towards apex. Hindwing cilia brownish, somewhat darker than those of forewing. Colour of dorsal side of male abdomen as ground colour of forewing; ventrally ivory; anal tuft covering segment VIII brownish ochre dorsally, ivory ventrally. Female abdomen about the same as in male but often a little richer in colour.

Diagnosis. *S.dissimilella* may easily be confused with *vartianae* Kasy (not in this work) and may only be separated from it with certainty by examination of the genitalia (see Kasy, 1962). Other species with a median white streak have more distinct borders in the streak or are not greyish brown.

Male genitalia (Fig. 97). Similar to those of *restigerella* (Z.) but differ in following: gnathos longer and slenderer, with four apical thornlike processes; aedeagus longer, hardly swollen at base; rectangular plates of sternum VIII with straight posterior margins; tergum VIII very deeply furcate.

Female genitalia (Fig. 132). Lamella antevaginalis two symmetrical, weak lobes; ductus bursae distinctly and uniformly sclerotized, transversely wrinkled, in posterior quarter abruptly widening, not gradually widening throughout its length as in *restigerella*, which has a smooth ductus bursae.

Distribution. Not in Denmark or Fennoscandia. Found in C and S Europe; occurs nearest to Scandinavia in Poland (Romaniszyn *in* Schille, 1930), C FGR and GDR (in Thuringia) (Rapp, 1936) and Belgium (Coenen, in litt.).

Biology. Larva undescribed. It feeds on *Helianthemum* in May and June and makes a thin silken tube below the basal leaves (Schütze, 1931). Adults fly from the end of June to the end of August, most frequently at the end of July.

32. *Scythris fuscoaenea* (Haworth, 1828)*
Figs. 12, 59, 98, 133.

Porrectaria fuscoaenea Haworth, 1828: 537.
Butalis schneideri Zeller, 1855: 194.

13-18 mm. Head, labial palpi, collar, tegulae, thorax and forewing dark bronze, most often with a greenish tinge but sometimes deeper brownish bronze. Whitish scales above eye and on scape. Forewing with pale brownish yellow scales on darker ground, especially around fold; apex with faint violet gloss. Hindwing 0.8 times as broad as forewing, dark brown, sparser scaled towards base. Cilia of forewing and hindwing brownish. Male abdomen fuscous above, faintly greenish; underside dirty ivory-yellow except for two posterior segments which are of same colour as upper side of abdomen. Female abdomen fuscous dorsally, ivory-coloured beneath.

Diagnosis. *S.fuscoaenea* is similar to *seliniella* (Z.), *clavella* (Z.) and *subseliniella* (Hein.) but has a pale abdominal underside, a narrower hindwing, more glossy forewing, whitish scales above the eye and on the scape and a rounded anal brush. *S.falla-cella* (Schl.) has a very stout abdomen which is brownish ventrally. *S.productella* (Z.) and *obscurella* (Scop.) are larger. Several other S and E European species may be confused with *fuscoaenea*, e.g. *podoliensis* Rbl., *flavilaterella* (Fuchs), *flaviventrella* (HS.), *speyeri* (Geyer), *amphonycella* (Geyer), etc.

Male genitalia (Fig. 98). Slightly asymmetrical and rather variable. Uncus a small rectangular plate. Gnathos long, slender and curved, strongly sclerotized terminally, cleft at apex. Aedeagus long, near cylindrical, bent just after middle and near apex. Tergum VIII broadly divided into two obliquely out-turned, digitate projections. Sternum VIII similarly bifurcate but each branch considerably broader and with bilobed apices.

Female genitalia (Fig. 133). Lamella postvaginalis weakly sclerotized, wrinkled medially; anteriorly with two small plates; antrum distinct; sternum VII with large anterior, elliptic area without sclerotization.

Distribution. Found in a few districts in SE Sweden. Not in Norway, Finland or Denmark. Otherwise in western C Europe, Italy and Greece. Also in southern half of Britain.

Biology. According to Benander (1965) the larva is found on *Helianthemum* in June in silken galleries amongst moss; slight webbing extends up the stems of the *Helianthemum*. The presence of the larva is easily noticed by the cream blotches on the leaves which wither and readily fall (Hofmann, 1893). The larva climbs in the thin web to different parts of the hostplant and, if disturbed, rapidly rushes backwards to the ground where it hides under dry leaves. It is then difficult to find owing to its good camouflage. It is brownish or greenish grey with yellow dorsal, dorsolateral, supraspiracular and (broad) subspiracular lines. The head and the prothoracic shield are yellow, each with two darker spots; the legs are yellow. Adults appear throughout the whole of July to mid-August.

33. *Scythris grandipennis* (Haworth, 1828)
Figs. 60, 99, 134.

Porrectaria grandipennis Haworth, 1828: 536.
Oecophora herbosella Herrich-Schäffer, 1855: 266.
Scythris cuencella Rebel, 1900: 175.
Scythris gallicella var. *unicolorella* de Joannis, 1909: 89.

14-20 mm. Head, labial palpi, antennae, collar, tegulae, thorax and forewing bronzy. Scales of forewing dark-tipped; sometimes medial and postmedial whitish scales (occasionally forming a medial streak), especially in female. Hindwing fuscous, 0.9 times as broad as forewing. Cilia of forewing and hindwing fuscous; at forewing termen with pale ochreous scales at base. Male abdomen proportionally thick, almost of same colour as forewing. Anal tuft curly, pale ochreous, pale terminally. Female abdomen coloured nearly as in male; last two segments with characteristic, dorsal depression.

Diagnosis. *S.grandipennis* may be confused with other large, weakly glossy species. *S.productella* (Z.) has a narrower hindwing and no whitish scales in the forewing; the anal tuft is not curly and is fuscous, not ochreous. *S.ericetella* (Hein.) is usually smaller and has browner forewings; the hindwing is narrower and, near the apex, browner. Other C and N European *Scythris* species are generally considerably smaller. In S Europe *grandipennis* is often confused with *cupreella* (Stgr.) and *scipionella* (Stgr.).

Continued on p. 122

Review of figures:
Nos. 23 - 62: Colour plates. Specimens approx. 5x.
Nos. 63 - 101: Male genitalia.
Nos. 102 - 136: Female genitalia.

Abbreviations applied to slides used for figures:
BM: British Museum (Natural History), London, England.
BÅB: Bengt Å. Bengtsson, Löttorp, Sweden.
JÄ: Eberhard Jäckh, Hörmanshofen, Germany.
MW: Naturhistorisches Museum, Wien, Austria.
NLW: Niels L. Wolff, Zoological Museum, Copenhagen, Denmark.

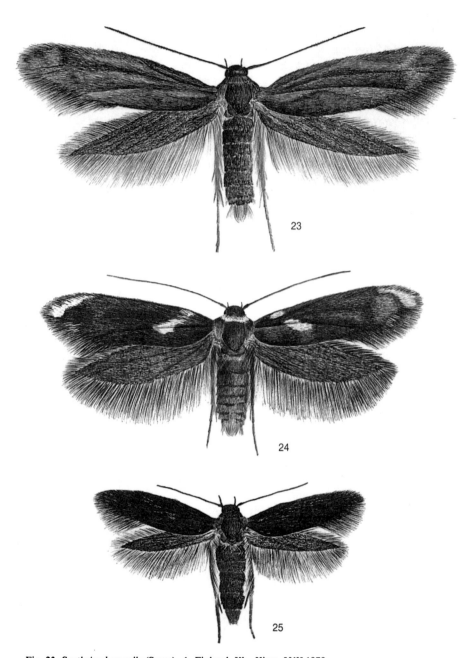

Fig. 23. *Scythris obscurella* (Scop.), ♂, Finland, Kb., Kitee, 5.VII.1978.
Fig. 24. *Scythris cuspidella* (Den. & Schiff.), ♂, Bulgaria, Predel, 15.VII.1979.
Fig. 25. *Scythris potentillella* (Z.), ♂, Sweden, Hall., Mästocka, 16.VI.1979.

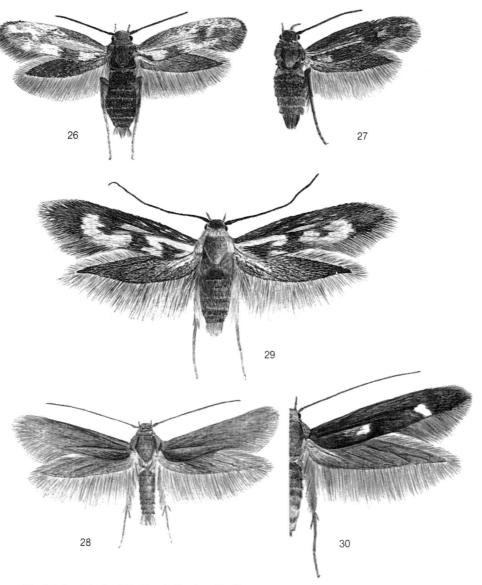

Fig. 26. *Scythris cicadella* (Z.), ♂, Sweden, Sk., Ö.Tvet, 16.VI.1978.
Fig. 27. *Scythris cicadella* (Z.), ♀, Sweden, Öl., Byrum, 10.VII.1977.
Fig. 28. *Scythris bifissella* (Hofm.), ♂, locality name illegible, 15.IV.1893, coll. Naturhist. Riksmus., Stockholm.
Fig. 29. *Scythris limbella* (F.), ♂, Sweden, Öl., Högby, 4.VIII.1976.
Fig. 30. *Scythris limbella* f. *obscura* (Stgr.), ♀, Sweden, Öl., Vedborm, 1.VIII.1976.

71

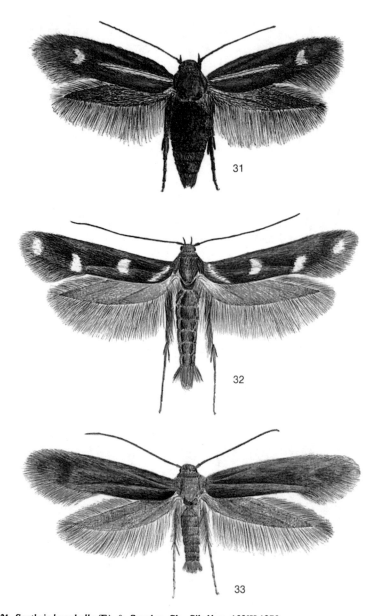

Fig. 31. *Scythris knochella* (F.), ♀, Sweden, Sk., Silvåkra, 15.VII.1979.
Fig. 32. *Scythris scopolella* (L.), ♂, Italy, Piemonte v. Susa, Racciamelone 1000 m, 7.VII.1979.
Fig. 33. *Scythris scopolella* (L.), ♂, Austria, Wien Umgebung (no date), coll. Naturhist. Riksmus., Stockholm.

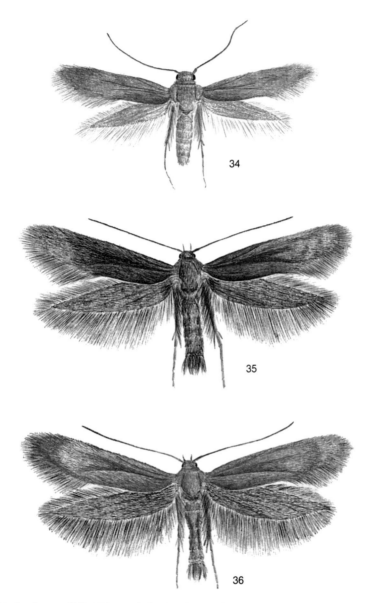

Fig. 34. *Scythris paullella* (HS.), ♀, Poland, Tolytsich, 23.VI.1898, coll. Naturhist. Riksmus., Stockholm.
Fig. 35. *Scythris clavella* (Z.), ♂, Poland, Toruń, 25.V.1975.
Fig. 36. *Scythris seliniella* (Z.), ♂, Latvia, Pêrkone, 17.VI.1979.

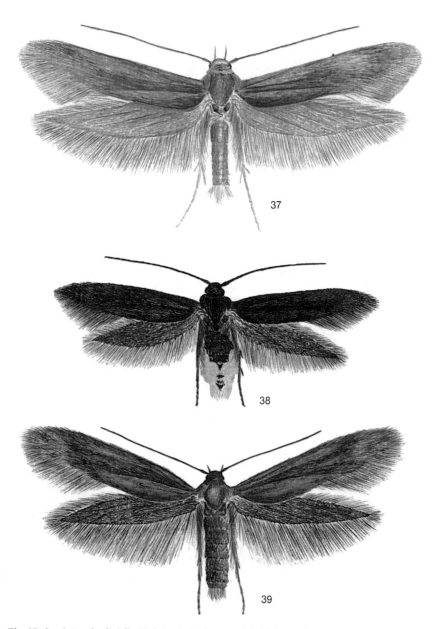

Fig. 37. *Scythris subseliniella* (Hein.), ♂, Spain, prov. Madrid, Aranjuez, 11.VI.1982.
Fig. 38. *Scythris sinensis* (Feld. & Rog.), ♂, Lithuania, Vilnius, 17.VII.1979.
Fig. 39. *Scythris productella* (Z.), ♂, Sweden, Nb., Seskarö, 6.VII.1975.

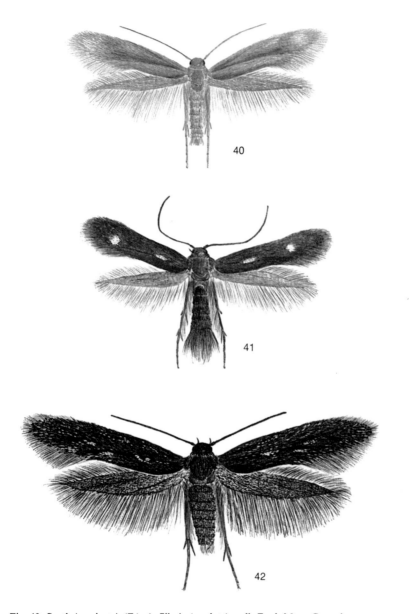

Fig. 40. *Scythris palustris* (Z.), ♂, Silesia (no date), coll. Zool. Mus., Copenhagen.
Fig. 41. *Scythris muelleri* (Mann), ♂, BRD, Bavaria mer., Carchinger Heide, 19.VI.1942.
Fig. 42. *Scythris inspersella* (Hb.), ♂, Sweden, Sm., Bäckebo, e.l. 26.VI.1979.

Fig. 43. *Scythris noricella* (Z.), ♂, Finland, Esbo, 24.VII.1926.
Fig. 44. *Scythris noricella* ssp. *latifoliella* Wolff, ♂, Greenland, Kapisigdlit, 19.VII.1950.
Fig. 45. *Scythris empetrella* Karsh. & Niel., ♂, Sweden, Sk., Viken, 9.VI.1980.
Fig. 46. *Scythris siccella* (Z.), ♂, Sweden, Sk., Ö.Tvet, 16.VI.1978.
Fig. 47. *Scythris tributella* (Z.), ♂, Italy, Liguria SW, Andora-Conna, a.l. 'Copa Conna', 10.IX.1979.

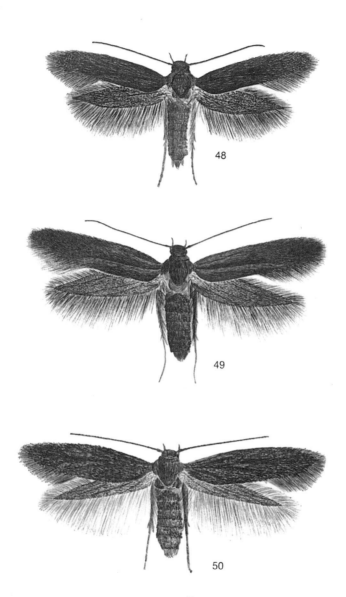

Fig. 48. *Scythris picaepennis* (Hw.), ♂, Sweden, Öl., Alvedsjöbodar, 19.VI.1978.
Fig. 49. *Scythris disparella* (Tgstr.), ♂, Sweden, Sm., Malghult, 6.VI.1979.
Fig. 50. *Scythris fuscopterella* Bengts., ♂, Sweden, Nb., Båtskärsnäs, 30.VI.1976.

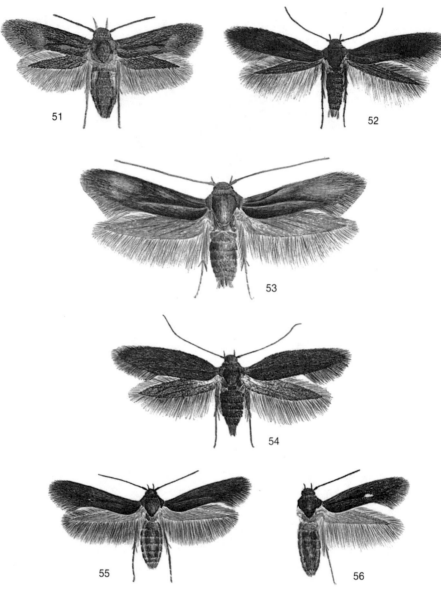

Fig. 51. *Scythris braschiella* (Hofm.), ♀, Germany, Mecklenburg (no date), coll. Naturhist. Riksmus., Stockholm.

Fig. 52. *Scythris laminella* (Den. & Schiff.), ♂, Sweden, Öl., Grönhögen, 15.VI.1977.

Fig. 53. *Scythris crassiuscula* (HS.), ♂, England, Kent, Trottiscliffe, 17.VI.1978, e.l. *Helianthemum nummularium.*

Fig. 54. *Scythris ericivorella* (Rag.), ♂, BRD, Schl.-Holst., Rendsburg, 20.VI.1978.

Fig. 55. *Scythris crypta* Hann., ♂, Sweden, Hall., Mästocka, 8.VI.1981.

Fig. 56. *Scythris crypta* Hann., ♂, Sweden, Hall., Mästocka, 8.VI.1981.

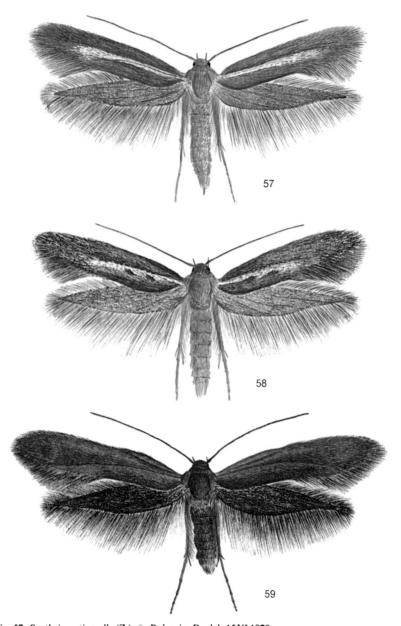

Fig. 57. *Scythris restigerella* (Z.), ♀, Bulgaria, Predel, 15.VI.1979.
Fig. 58. *Scythris dissimilella* (HS.), ♂, Italy, Vologno, Castelnovo né Monti (RE), 650 m, 31.VII.1976.
Fig. 59. *Scythris fuscoaenea* (Hw.), ♂, Sweden, Gtl., Hejnum, 11.VII.1980.

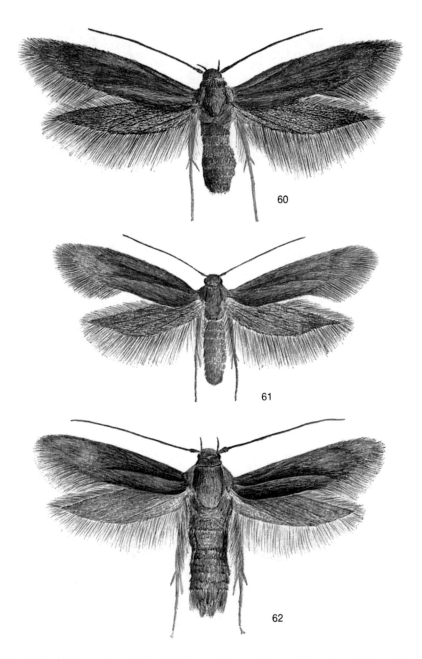

Fig. 60. *Scythris grandipennis* (Hw.), ♂, France, Bretagne, Vannes (no date), coll. Naturhist. Riksmus., Stockholm.

Fig. 61. *Scythris ericetella* (Hein.), ♂, Germany, Nassau (no date), coll. Zool. Mus., Copenhagen.

Fig. 62. *Scythris fallacella* (Schl.), ♂, Italy, Liguria, C.della Melosa, (IM), 1600 m, 29.VI.1971.

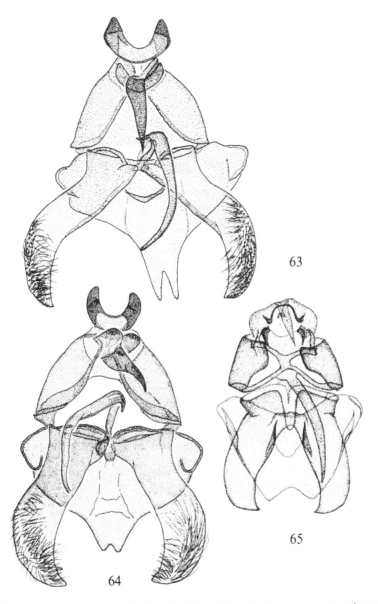

Fig. 63. *Scythris obscurella* (Scop.), Finland, Kirjavalhti (no date), male genitalia (BÅB 1256).
Fig. 64. *Scythris cuspidella* (Den. & Schiff.), Bulgaria, Predel, 17.VII.1979, male genitalia (BÅB 1482).
Fig. 65. *Scythris potentillella* (Z.), Sweden, Sk., S.Stenshuvud, 14.VIII.1976, male genitalia (BÅB 707).

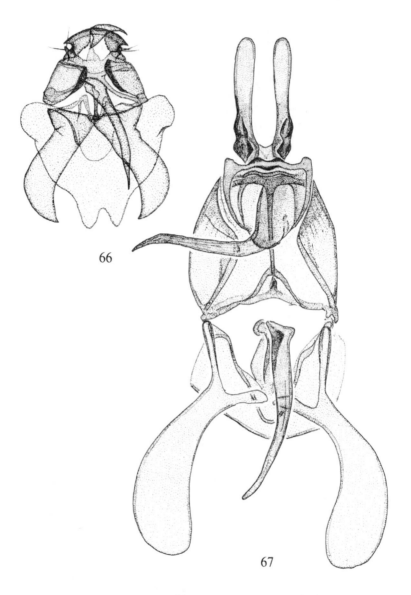

Fig. 66. *Scythris cicadella* (Z.), Sweden, Öl., Byrum, 10.VII.1977, male genitalia (BÅB 711).
Fig. 67. *Scythris bifissella* (Hofm.), Austria, Burgenland, Illmitz am Neusiedlersee, 19.VIII.1961, male genitalia (MW 4205).

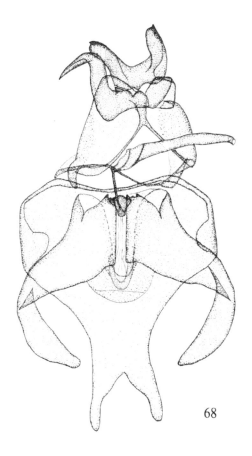

68

Fig. 68. *Scythris limbella* (F.), Sweden, Öl., Högby, 4.VIII.1976, male genitalia (BÅB 713).

69

Fig. 69. *Scythris knochella* (F.), Sweden, Sk., Silvåkra, 14.VIII.1976, male genitalia (BÅB 533).

70

Fig. 70. *Scythris scopolella* (L.), Austria, Wien Umgeb. (no date), coll. Naturhist. Riksmus., Stockholm, male genitalia (BÅB 0082X).

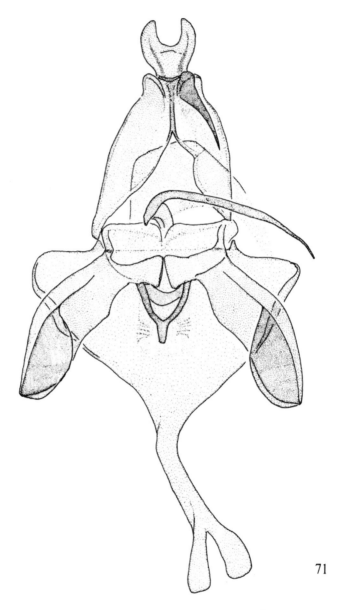

71

Fig. 71. *Scythris paullella* (HS.), Bohemia, Reichst. Mn, coll. Mus. Wien, male genitalia (JÄ 8345).

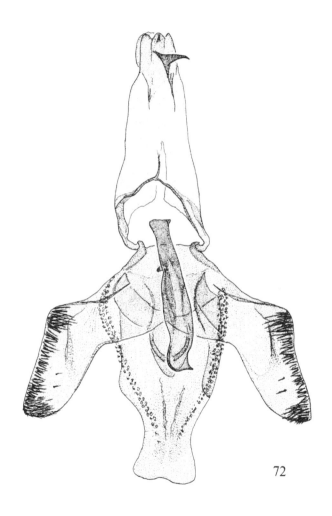

72

Fig. 72. *Scythris clavella* (Z.), Poland, Poznan, Glówinice poligon, 11.VI.1940, male genitalia (BÅB 0068X).

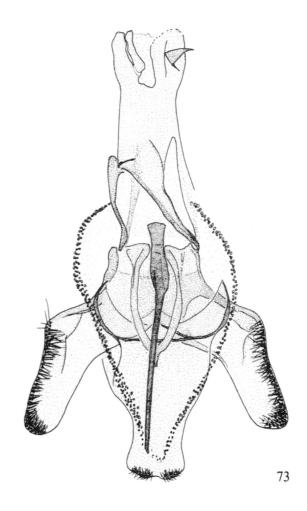

73

Fig. 73. *Scythris seliniella* (Z.), Latvia, Pêrkone, 17.VI.1979, male genitalia (BÅB 1246).

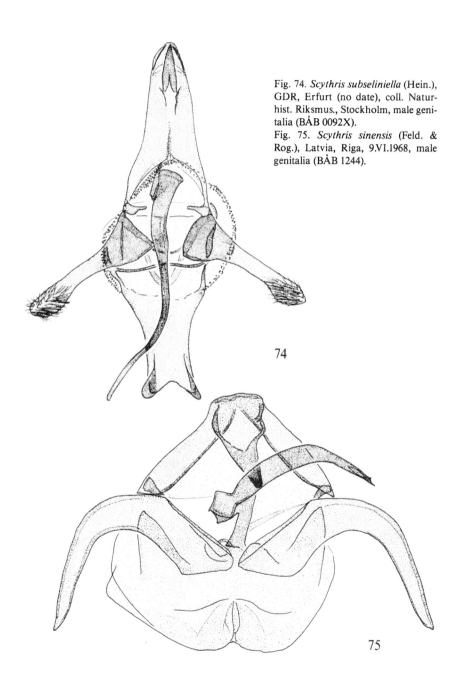

Fig. 74. *Scythris subseliniella* (Hein.), GDR, Erfurt (no date), coll. Naturhist. Riksmus., Stockholm, male genitalia (BÅB 0092X).
Fig. 75. *Scythris sinensis* (Feld. & Rog.), Latvia, Riga, 9.VI.1968, male genitalia (BÅB 1244).

74

75

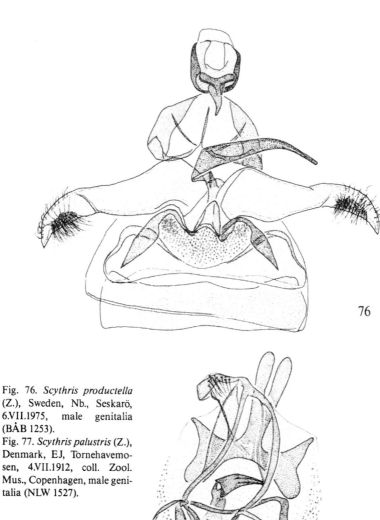

76

Fig. 76. *Scythris productella*
(Z.), Sweden, Nb., Seskarö,
6.VII.1975, male genitalia
(BÅB 1253).
Fig. 77. *Scythris palustris* (Z.),
Denmark, EJ, Tornehavemo-
sen, 4.VII.1912, coll. Zool.
Mus., Copenhagen, male geni-
talia (NLW 1527).

77

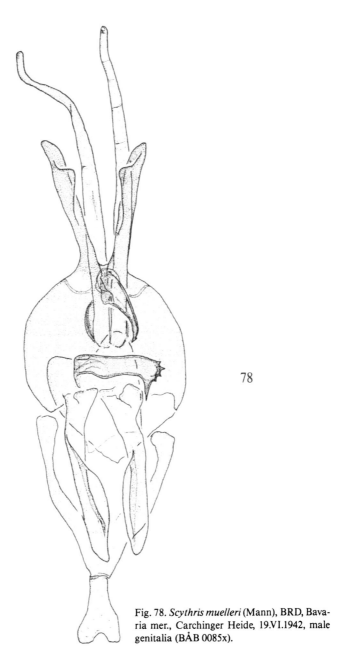

78

Fig. 78. *Scythris muelleri* (Mann), BRD, Bavaria mer., Carchinger Heide, 19.VI.1942, male genitalia (BÅB 0085x).

79   80

Fig. 79. *Scythris inspersella* (Hb.), Sweden, Sk., Dalby kronopark, e.l. 14.VIII.1974, male genitalia (BÅB 715).
Fig. 80. *Scythris noricella* (Z.), Greenland, Sarqaq, 29.VII.1949, male genitalia (BÅB 1249).

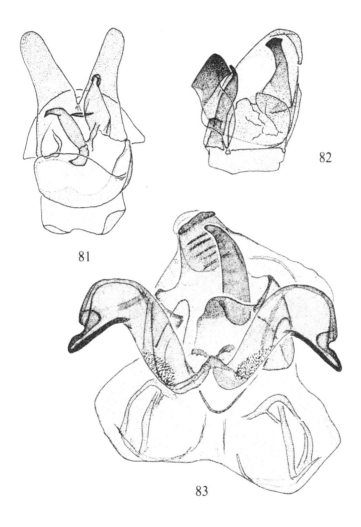

Fig. 81. *Scythris empetrella* Karsh. & Niel., Sweden, Sk., Sandhammaren, 10.VI.1980, male genitalia (BÅB 1250).
Fig. 82. *Scythris siccella* (Z.), Sweden, Sk., Vitemölla, 17.VI.1978, male genitalia (BÅB 743).
Fig. 83. *Scythris tributella* (Z.), Germany, West-Thüringen, Gr.-Hörselberg, 8.VII.1934, male genitalia (JÄ 5271).

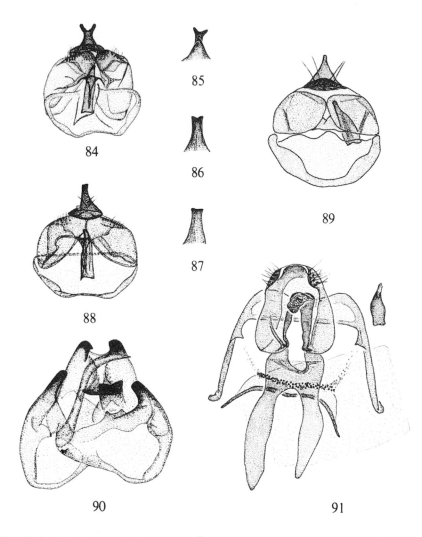

Fig. 84. *Scythris picaepennis* (Hw.), Sweden, Öl., Högby, 24.VI.1977, male genitalia (BÅB 706).
Figs. 85-87. *Scythris picaepennis* (Hw.). Variation in form of uncus. – 85: (BÅB 751); 86: (BÅB 754); 87: (BÅB 753).
Fig. 88. *Scythris picaepennis* (Hw.), Sweden, Öl., Högby, 23.VII.1976, male genitalia (BÅB 426).
Fig. 89. *Scythris disparella* (Tgstr.), Sweden, Ög., Omberg, 2.VII.1979, male genitalia (BÅB 909).
Fig. 90. *Scythris fuscopterella* Bengts., Sweden, Nb., Båtskärsnäs, 30.VI.1976, male genitalia (BÅB 427).
Fig. 91. *Scythris braschiella* (Hofm.), Germany, Potsdam XVIII, Z.2.VIII.1899, Stat.arm., coll. Naturhist. Riksmus., Stockholm, male genitalia (BÅB 0056X).

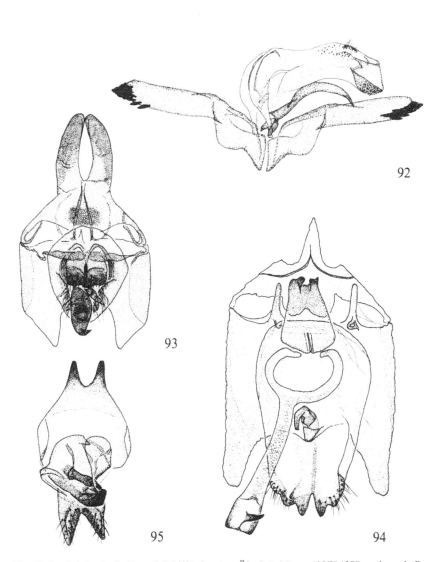

Fig. 92. *Scythris laminella* (Den. & Schiff.), Sweden, Öl., Grönhögen, 13.VII.1977, male genitalia (BÅB 619).

Fig. 93. *Scythris crassiuscula* (HS.), England, Kent, Trottiscliffe, 17.VI.1978, e.l. *Helianthemum nummularium,* male genitalia (BÅB 1469).

Fig. 94. *Scythris ericivorella* (Rag.), BRD, Schl.-Holst., Rendsburg, 20.VI.1978, male genitalia (BÅB 1251).

Fig. 95. *Scythris crypta* Hann., Sweden, Hall., Mästocka, 8.VI.1981, male genitalia (BÅB 1478).

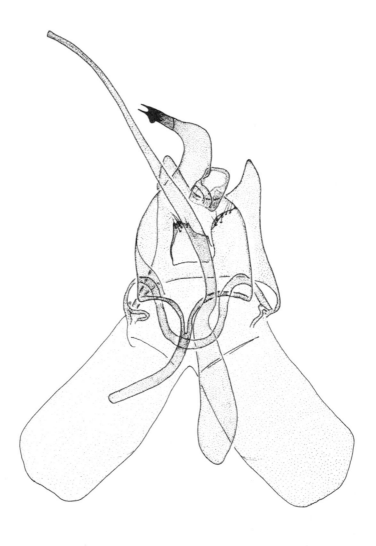

96

Fig. 96. *Scythris restigerella* (Z.), Bulgaria, Bazkovo, 30.VII.1979, male genitalia (BÅB 1247).

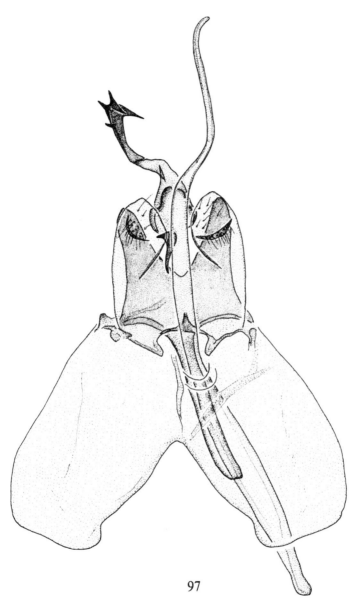

97

Fig. 97. *Scythris dissimilella* (HS.), Italy, Vologno, Castelnovo né Monti (RE), 650 m, 31.VII.1976, male genitalia (BÅB 1481).

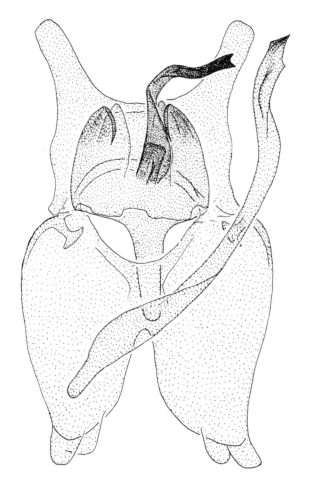

98

Fig. 98. *Scythris fuscoaenea* (Hw.), Sweden, Öl., Långe Erik, 26.VII.1976, male genitalia (BÅB 1255).

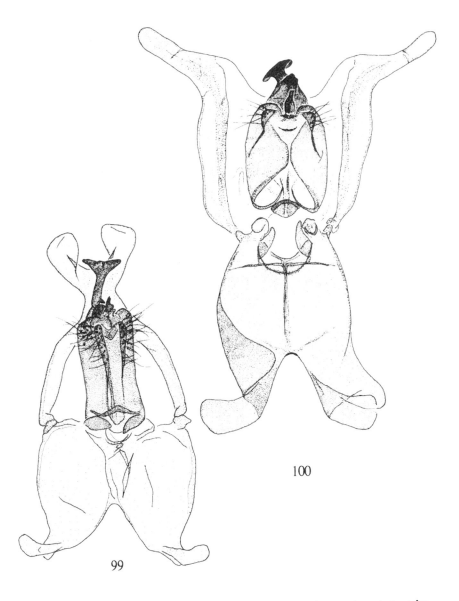

Fig. 99. *Scythris grandipennis* (Hw.), France, Bretagne, Vannes (no date), male genitalia (BÅB 1472).

Fig. 100. *Scythris ericetella* (Hein.), BRD, Nassau (no date), coll. Naturhist. Riksmus., Stockholm, male genitalia (BÅB 0042X).

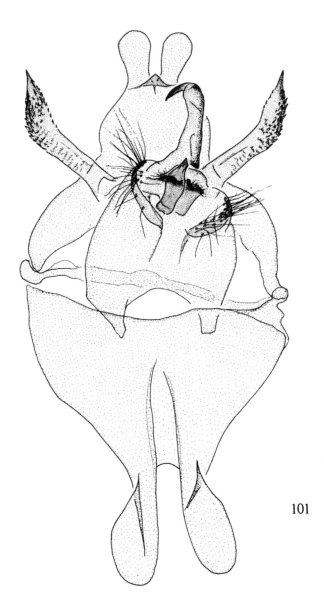

101

Fig. 101. *Scythris fallacella* (Schl.), England, Cumbria, Grange, 27.VI.1979, e.l. *Helianthemum,* male genitalia (BÅB 1474).

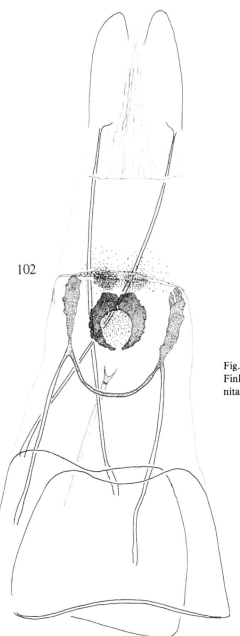

102

Fig. 102. *Scythris obscurella* (Scop.), Finland, Vaaseni (no date), female genitalia (BÅB 1257).

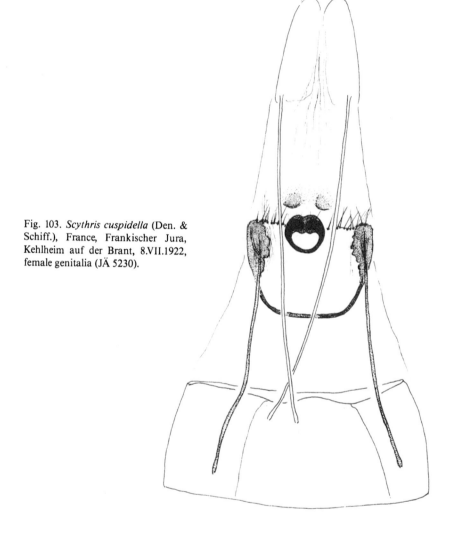

Fig. 103. *Scythris cuspidella* (Den. &
Schiff.), France, Frankischer Jura,
Kehlheim auf der Brant, 8.VII.1922,
female genitalia (JÄ 5230).

103

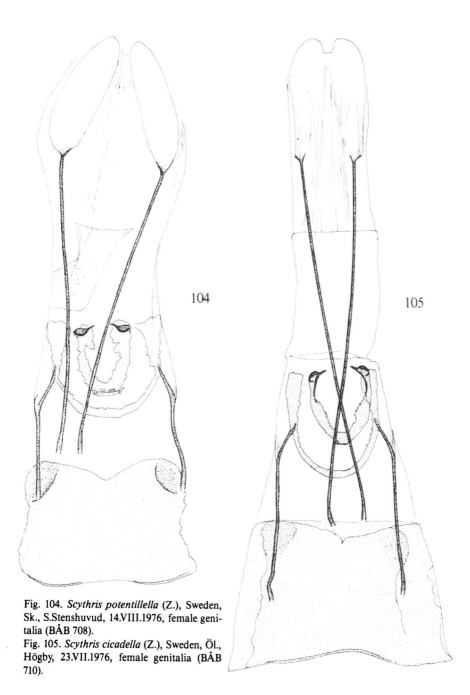

104

105

Fig. 104. *Scythris potentillella* (Z.), Sweden, Sk., S.Stenshuvud, 14.VIII.1976, female genitalia (BÅB 708).
Fig. 105. *Scythris cicadella* (Z.), Sweden, Öl., Högby, 23.VII.1976, female genitalia (BÅB 710).

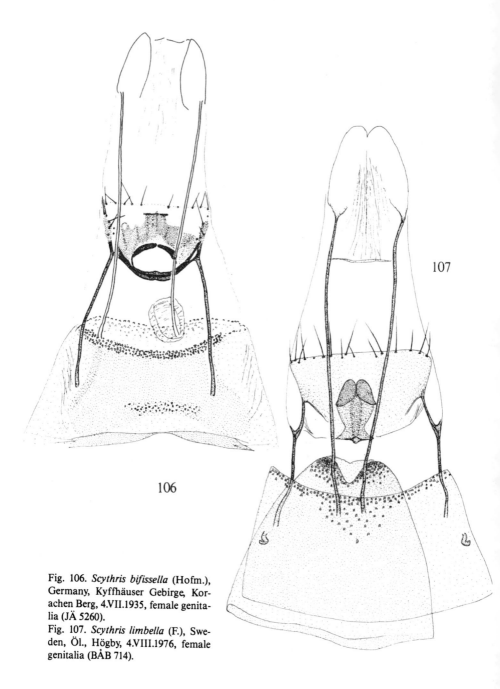

Fig. 106. *Scythris bifissella* (Hofm.), Germany, Kyffhäuser Gebirge, Korachen Berg, 4.VII.1935, female genitalia (JÄ 5260).

Fig. 107. *Scythris limbella* (F.), Sweden, Öl., Högby, 4.VIII.1976, female genitalia (BÅB 714).

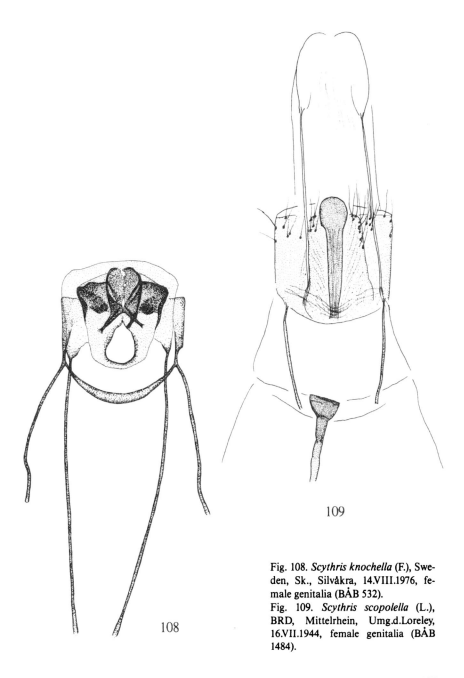

108

109

Fig. 108. *Scythris knochella* (F.), Sweden, Sk., Silvåkra, 14.VIII.1976, female genitalia (BÅB 532).

Fig. 109. *Scythris scopolella* (L.), BRD, Mittelrhein, Umg.d.Loreley, 16.VII.1944, female genitalia (BÅB 1484).

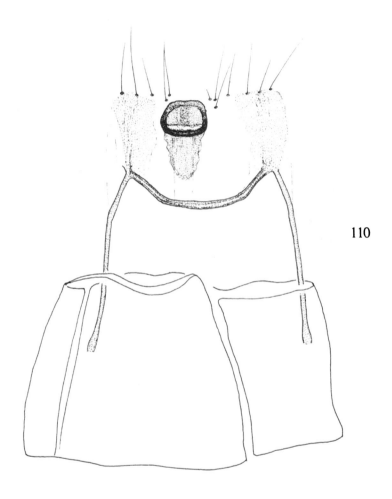

110

Fig. 110. *Scythris paullella* (HS.), Poland, Tolytsich, 23.VI.1898, coll. Naturhist. Riksmus., Stockholm, female genitalia (BÅB 0080X).

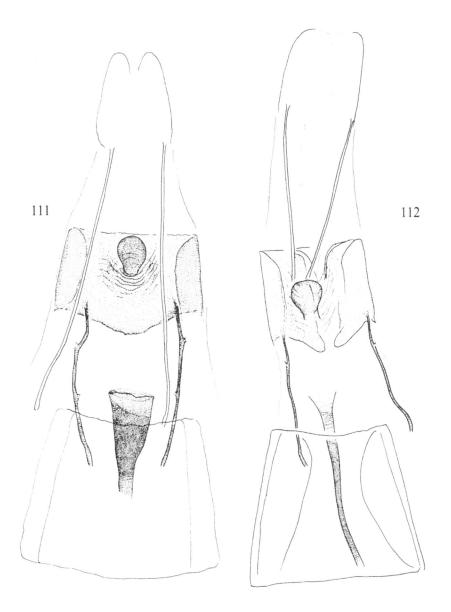

Fig. 111. *Scythris clavella* (Z.), Germany, Thüringen (no date), coll. Zool. Mus., Copenhagen, female genitalia (BÅB 0046X).

Fig. 112. *Scythris seliniella* (Z.), Austria, Niederösterreich, Oberweiden, Rennplatzwiesen, 9.VII.1980, female genitalia (BÅB 1485).

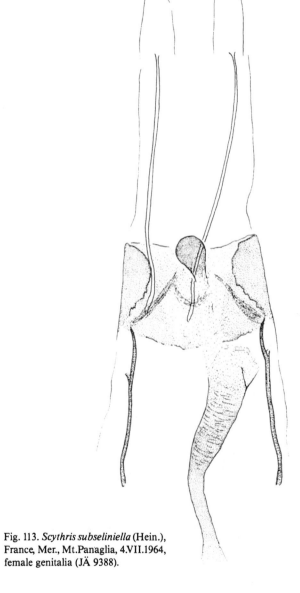

113

Fig. 113. *Scythris subseliniella* (Hein.),
France, Mer., Mt.Panaglia, 4.VII.1964,
female genitalia (JÄ 9388).

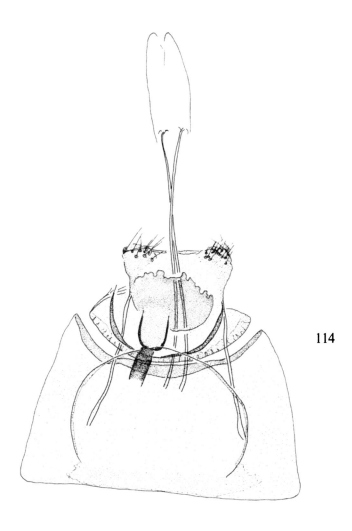

114

Fig. 114. *Scythris sinensis* (Feld. & Rog.), Lithuania, Vilnius, 5.VI.1973, female genitalia (BÅB 1245).

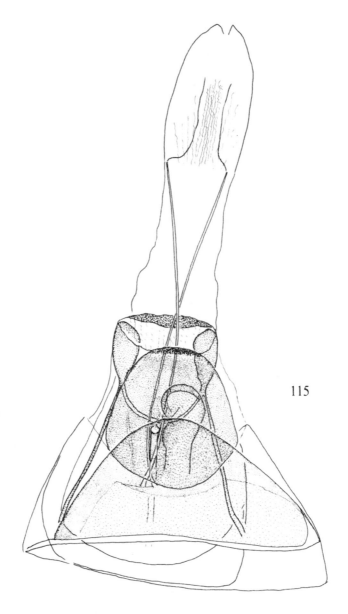

115

Fig. 115. *Scythris productella* (Z.), Sweden, Nb., Seskarö, 7.VII.1975, female genitalia (BÅB 1471).

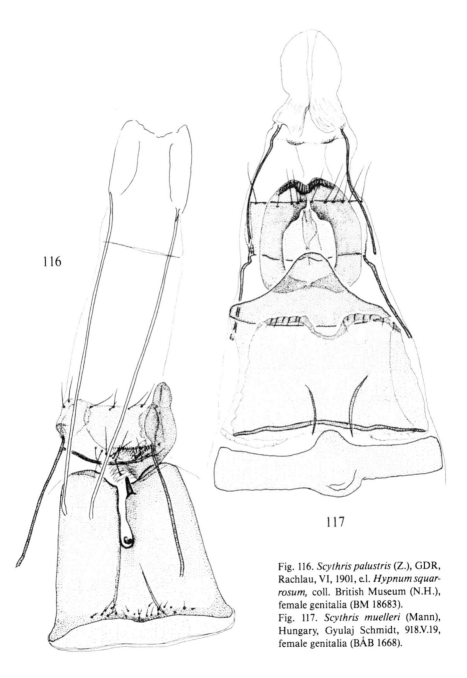

116

117

Fig. 116. *Scythris palustris* (Z.), GDR, Rachlau, VI, 1901, e.l. *Hypnum squarrosum,* coll. British Museum (N.H.), female genitalia (BM 18683).

Fig. 117. *Scythris muelleri* (Mann), Hungary, Gyulaj Schmidt, 918.V.19, female genitalia (BÅB 1668).

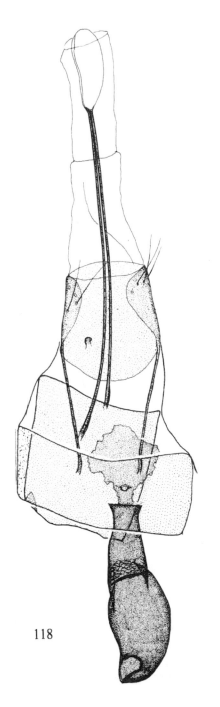

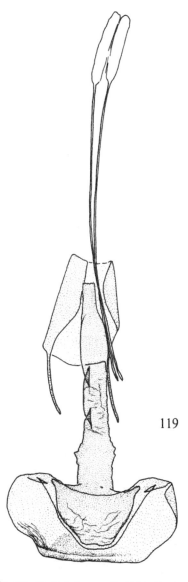

119

Fig. 118. *Scythris inspersella* (Hb.), Sweden, Nb., Notträsk, e.l. 2.VIII.1975, female genitalia (BÅB 716).

Fig. 119. *Scythris noricella* (Z.), Finland, Esbo, 24.VII.1926, female genitalia (BÅB 1248).

118

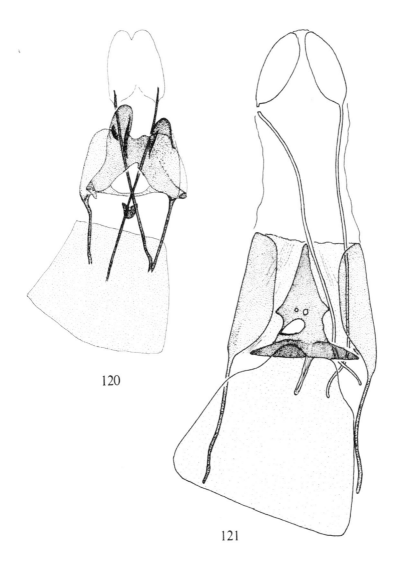

120

121

Fig. 120. *Scythris empetrella* Karsh. & Niel., Denmark, EJ, Anholt, 11.VII.1974, female genitalia (BÅB 717).
Fig. 121. *Scythris siccella* (Z.), Sweden, Sk., Vitemölla, 17.VI.1978, female genitalia (BÅB 742).

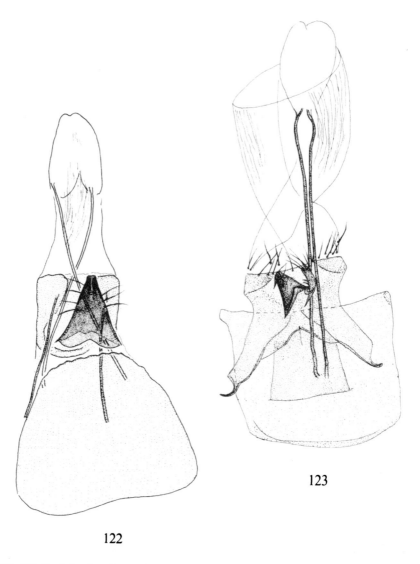

123

122

Fig. 122. *Scythris tributella* (Z.), BRD, Bornich, Rheinberge, 14.VIII.1901, coll. Naturhist. Riksmus., Stockholm, female genitalia (BÅB 0079X).

Fig. 123. *Scythris picaepennis* (Hw.), Sweden, Öl., Högby, 23.VII.1976, female genitalia (BÅB 705).

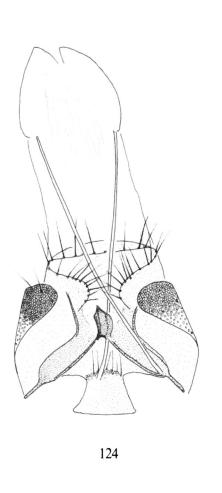

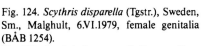

124

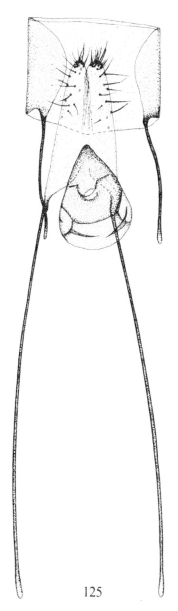

125

Fig. 124. *Scythris disparella* (Tgstr.), Sweden, Sm., Malghult, 6.VI.1979, female genitalia (BÅB 1254).
Fig. 125. *Scythris fuscopterella* Bengts., Sweden, Nb., Båtskärsnäs, 30.VI.1976, female genitalia (BÅB 545).

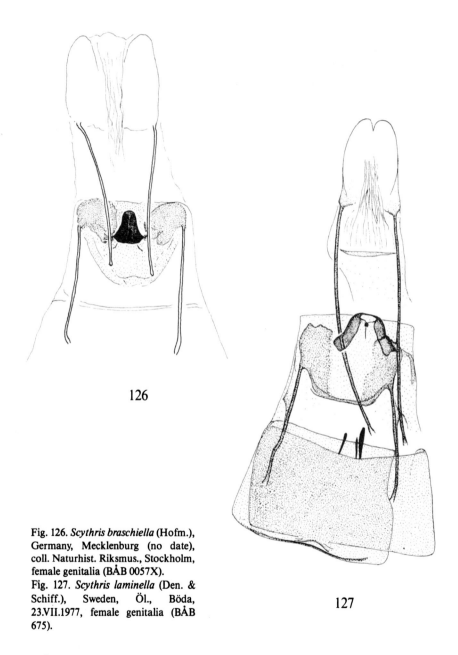

126

Fig. 126. *Scythris braschiella* (Hofm.), Germany, Mecklenburg (no date), coll. Naturhist. Riksmus., Stockholm, female genitalia (BÅB 0057X).
Fig. 127. *Scythris laminella* (Den. & Schiff.), Sweden, Öl., Böda, 23.VII.1977, female genitalia (BÅB 675).

127

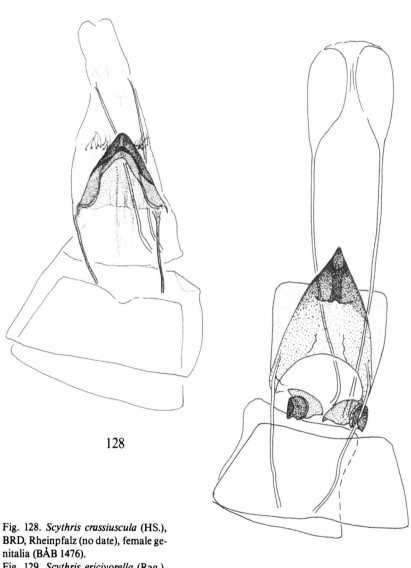

128

Fig. 128. *Scythris crassiuscula* (HS.), BRD, Rheinpfalz (no date), female genitalia (BÅB 1476).
Fig. 129. *Scythris ericivorella* (Rag.), BRD, Schl.-Holst., Rendsburg, 20.VI.1978, female genitalia (BÅB 1252).

129

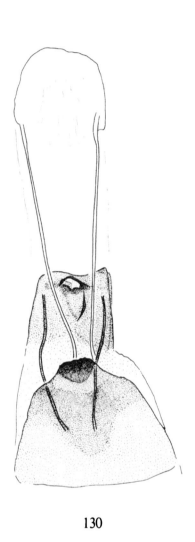

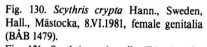

130

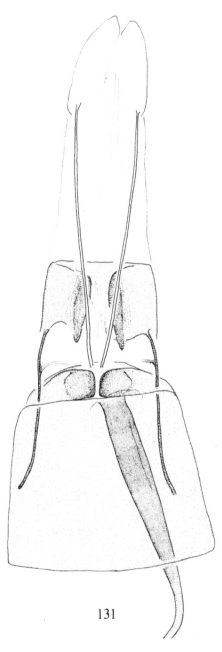

Fig. 130. *Scythris crypta* Hann., Sweden, Hall., Mästocka, 8.VI.1981, female genitalia (BÅB 1479).
Fig. 131. *Scythris restigerella* (Z.), Austria, Wien, Vamler?, 3.VIII.1902, coll. Naturhist. Riksmus., Stockholm, female genitalia (BÅB 0088X).

131

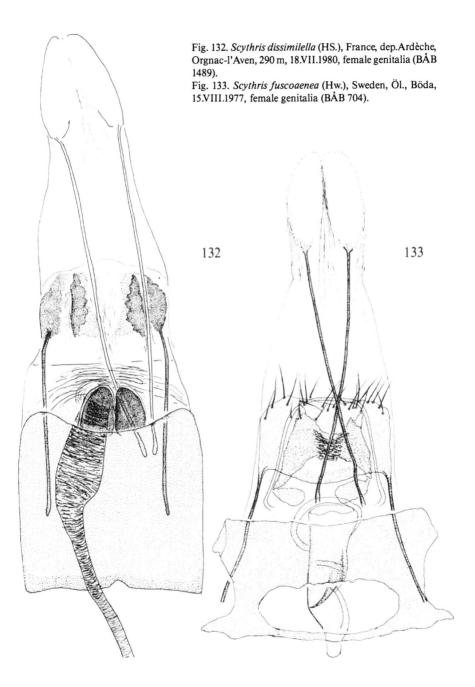

Fig. 132. *Scythris dissimilella* (HS.), France, dep.Ardèche, Orgnac-l'Aven, 290 m, 18.VII.1980, female genitalia (BÅB 1489).
Fig. 133. *Scythris fuscoaenea* (Hw.), Sweden, Öl., Böda, 15.VIII.1977, female genitalia (BÅB 704).

132

133

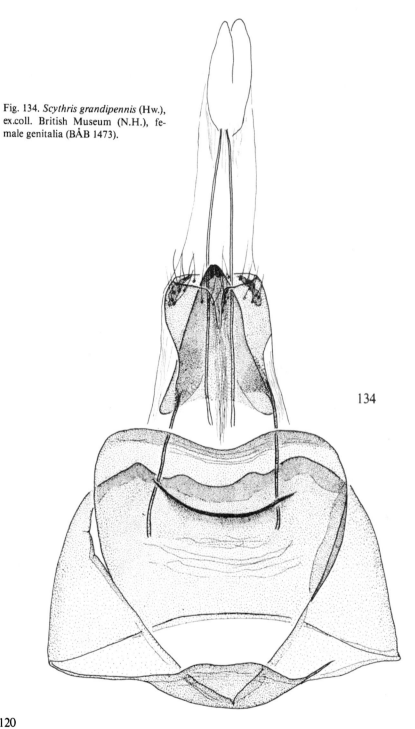

Fig. 134. *Scythris grandipennis* (Hw.), ex.coll. British Museum (N.H.), female genitalia (BÅB 1473).

134

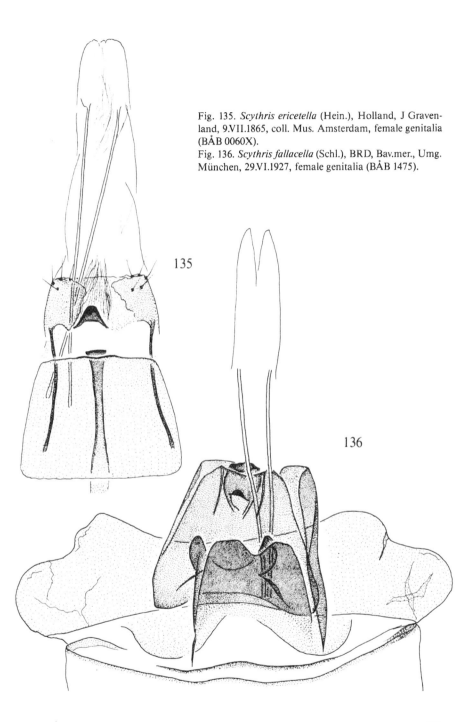

Fig. 135. *Scythris ericetella* (Hein.), Holland, J Graven-
land, 9.VII.1865, coll. Mus. Amsterdam, female genitalia
(BÅB 0060X).
Fig. 136. *Scythris fallacella* (Schl.), BRD, Bav.mer., Umg.
München, 29.VI.1927, female genitalia (BÅB 1475).

135

136

Male genitalia (Fig. 99). Valva a small, setose plate. Gnathos at apex with lateral projections and minute central tooth; near base a small projection. Aedeagus small, pear-shaped. Tergum VIII symmetrically bilobed distally. Sternum VIII bisected, united medially, strongly diverging posteriorly, terminating in small, narrow lobes.

Female genitalia (Fig. 134). Lamella postvaginalis triangular with acute-angled basal emargination. Segment VIII long, laterally concave. Sternum VII and tergum VII with concave, posterior margins, posterior margin of sternum VII slightly scalloped.

Distribution. Not in Denmark or Fennoscandia. Southern part of Britain, E Ireland (O'Connor, in litt.), western C Europe and the Iberian peninsula.

Biology. Larva on *Ulex* and *Genista*, in Britain preferring *Ulex minor* Roth. (Emmet, 1979). It is dark greyish green with a brownish head, pale green dorsal line and dark green subdorsal line. It makes an extensive web amongst fresh shoots and incorporates excrement into the web (Meyrick, 1928; Roche, in litt.). Pupation takes place in a rather thin web near the stem or between tight leaves. Adults emerge in June. Locally abundant.

Note. Caradja (1920) described var. *porrectella* which was said to have a narrower forewing and to lack pale scales. He also pointed out certain divergences between *cuencella* (Rbl.) and *grandipennis*. It is well known that specimens of *grandipennis* and also of other species (e.g. *subseliniella* (Hein.)) from the Iberian peninsula possess an increased number of whitish scales in the forewing. Jäckh (1977) remarks that specimens of *grandipennis* from Spain are larger while they are comparatively small in western France.

34. *Scythris ericetella* (Heinemann, 1872)
Figs. 61, 100, 135.

Butalis ericetella Heinemann, 1872: 280.

13-17 mm. Head, labial palpi, antennae, collar, tegulae, thorax and forewing olive-brown with very faint greenish tone. Palpi sometimes with dorsal whitish scales. Forewing glossy, with unicolorous scales, in some cases with paler scales at termen and in preapical area. Hindwing 0.8 times as broad as forewing, brownish with faint purplish gloss, richer in colour at apex. Cilia of forewing and hindwing fuscous, with slight purplish tinge, especially in hindwing. Male abdomen slender, greyish ochreous, ventrally paler – dirty ivory. Anal tuft curled and of almost same colour as preceding segments. Female abdomen brown or greyish ochreous, posteriorly paler – pale ochreous. Ventral side ivory, anteriorly somewhat darker.

Diagnosis. Similar to *grandipennis* (Hw.) which is usually larger; the abdomen in *grandipennis* is ventrally darker. *S.seliniella* (Z.), *clavella* (Z.) and *subseliniella* (Hein.) all have a slenderer, darker abdomen and (in males) a trifurcate anal brush. *S.fallacella* (Schl.) is darker and has a thicker abdomen with a medial, posterior depression. While habitat may give an indication of identity, only genital dissection will provide a positive identification of this species.

Male genitalia (Fig. 100). Valva a small lobe with many fine bristles. Gnathos anvil-shaped. Aedeagus short, tubular. Tergum VIII variable, posteriorly with two divergent arms (cf. Jäckh, 1977). Sternum VIII furcate, posterior lobes of varying breadth and length.

Female genitalia (Fig. 135). Lamella postvaginalis small, trapezoidal, anteriorly concave, lateral margins with strong sclerotization. Ductus bursae well-defined, widening posteriorly and anteriorly.

Distribution. No records in Denmark or Fennoscandia. Otherwise reported from the Netherlands, Belgium, France, Austria and Spain (Jäckh, 1977). On heaths.

Biology. The larva is found on *Calluna vulgaris* (L.) Hull. and on different *Erica* species (Jäckh, 1977) but appears to be undescribed. Adults fly apparently in two broods, from the end of May to early September. In the Netherlands they appear from the end of June to early August, judging from museum materials.

Note. Gerasimov (1930) described *Scythris lativittis* from a male and a female from C Asia. Jäckh (1977) has suggested that *lativittis* is possibly a synonym of *ericetella*. There are some details, however, which indicate that it may be a separate species. The forewing markings seem to be different from those of *ericetella*. The uncus is furcate, not a small flap; the gnathos is fairly long and hooked, not short and anvil-shaped; tergum VIII is strongly asymmetrical and there are other differences. Further study of the status of *lativittis* is required.

35. *Scythris fallacella* (Schläger, 1847)
Figs. 62, 101, 136.

Oecophora fallacella Schläger, 1847: 238.
Oecophora armatella Herrich-Schäffer, 1855: 267; Zeller, 1855:184.
Butalis hydrargyrella Steudel & Hofmann, 1882: 223; Hannemann, 1958b: 86.

12-15 mm. Head, labial palpi, antennae, collar, tegulae, thorax and forewing greyish olive-brown with a greenish or yellowish tinge. Scales smooth, rather narrow, concolorous and glossy. Hindwing 0.9 times as broad as forewing, pale fuscous. Cilia of forewing and hindwing fuscous, in hindwing with faint cilia line. Male abdomen very stout, uniformly olive-brown, almost of same colour as in forewing. Anal tuft brown ochreous, compact, obtusely pointed. Dorsal surface of female abdomen ivory to ochreous posteriorly and anteriorly, sometimes brownish or dark ochreous medially. Ventral surface ochreous brown anteriorly, last two or three segments whitish or ivory. Apex with distinct, medial depression; lateral lobes of last two segments conspicuous.

Diagnosis. The very stout abdomen distinguishes *fallacella* from *fuscoaenea* (Hw.), *clavella* (Z.), *seliniella* (Z.) and *subseliniella* (Hein.). Furthermore, the antennae in *fallacella* are only ciliated anteriorly. Other species without a red tinge in the apical area are considerably smaller. *S.fallacella* is similar to some other European species with broad abdomen (e.g. *Scythris baldensis* Pass. d'Ent.) and can only be safely identified by dissection of the genitalia.

Male genitalia (Fig. 101). Valva very small, a bent, setose fold. Uncus rudimentary; gnathos a stout, hooked prong, apex pointed; aedeagus small, bottle-shaped; tergum VIII apically deeply cleft, at midlength with conspicuous pair of stout, dentate and pointed processes; sternum VIII with two symmetrical posterior lobes.

Female genitalia (Fig. 136) Lamella postvaginalis large, trapezoidal. Segment VIII with very small apophyses anteriores. Tergum VII predominant, posteriorly concave, with lateral, upturned lobes. Interpretation of other components very difficult.

Distribution. Not found in Denmark or Fennoscandia. In Britain in a few places near Lancaster (Pierce & Metcalfe, 1935a; MacKechnie Jarvis, 1975). C and SE Europe. Recorded closest to the Scandinavian area at Kyffhäuser in GDR. Records from S France and Portugal are doubtful (Jäckh, unpubl.).

Biology. Larva undescribed. It feeds on various *Helianthemum* spp., and, according to Schütze (1931), produces a web covering almost all of the plant. Pupation takes place on the ground. In Britain adults appear in May and June and are seen in evening sunshine (Emmet, 1979). Adults are also found in August, which indicates that there is a second brood (Emmet, loc.cit.).

Note. '*Scythris alboalvella* (FR.)' is a manuscript name and regarded as a synonym of *fallacella* by both Zeller (1855) and Caradja (1920). '*S.albiventella* (Mann)' (?), also a manuscript name, was regarded as conspecific with *fallacella* by Caradja (1920).

Appendix

Scythris gracilella (Heinemann, 1877)

Butalis gracilella Heinemann, 1877: 452.

The original (translated) description reads as follows: 'Small, with short, slightly ascending labial palpi. Forewing elongately pointed, dark olive-green with purple tone, dull glossy. Hindwing 0.4 times as broad as forewing, slender, blackish grey. Male abdomen very slender, blackish, with long, oval anal brush. Forewing 2.25 lines, expansion 4.5 lines (5.7 mm and 11.4 respectively).

Close to *parvella*, but darker. Forewing somewhat broader, dark olive-green, without yellow tinge, faintly purplish. Hindwing shaped as in *parvella*, slightly narrower, but darker. Labial palpi less ascending. Male abdomen very slender, almost black, ventrally paler; anal brush as dark as in *parvella* but longer and denser, laterally with some divergent but incurved hairs. I do not know the female. In Sachsen, in June.'

It has not been possible to obtain further information on this taxon. Attempts to trace the type material have hitherto been unsuccessful. The description is passably applicable to several species with narrow hindwings, such as *laminella* (Den. & Schiff.), *tributella* (Z.), etc. Without original material, the identity of this species must remain in doubt.

				DENMARK												
		Germany	G. Britain	SJ	EJ	WJ	NWJ	NEJ	F	LFM	SZ	NWZ	NEZ	B	Sk.	Bl.
Scythris obscurella (Scop.)	1	●														
S. cuspidella (Den. & Schiff.)	2	●														
S. potentillella (Z.)	3	●	●	●	●			●				●	●	●	●	●
S. cicadella (Z.)	4	●	●			●	●	●	●	●					●	●
S. bifissella (Hofm.)	5	●														
S. limbella (F.)	6	●	●			●			●			●	●	●	●	●
S. knochella (F.)	7	●										●	●			
S. scopolella (L.)	8	●														
S. paulella (HS.)	9	●														
S. clavella (Z.)	10	●														
S. seliniella (Z.)	11	●														
S. subseliniella (Hein.)	12	●														
S. sinensis (Feld. & Rog.)	13															
S. productella (Z.)	14	●														
S. palustris (Z.)	15	●			●				●							
S. muelleri (Mann)	16	●														
S. inspersella (Hb.)	17	●		●	●	●	●	●	●	●	●	●	●	●	●	●
S. noricella (Z.)	18	●														
S. empetrella Karsh. & Niel.	19		●			●	●	●	●			●	●	●		
S. siccella (Z.)	20	●	●			●	●	●	●						●	●
S. tributella (Z.)	21	●														
S. picaepennis (Hw.)	22	●	●			●	●	●	●	●	●				●	●
S. disparella (Tgstr.)	23	●														●
S. fuscopterella Bengts.	24															
S. braschiella (Hofm.)	25	●														
S. laminella (Den. & Schiff.)	26	●												●		●	●
S. crassiuscula (HS.)	27	●	●														
S. ericivorella (Rag.)	28	●				●	●	●	●	●						
S. crypta Hann.	29															
S. restigerella (Z.)	30	●														
S. dissimilella (HS.)	31	●														
S. fuscoaenea (Hw.)	32	●	●														
S. grandipennis (Hw.)	33	●	●														
S. ericetella (Hein.)	34	●	●														
S. fallacella (Schl.)	35	●														

	Hall.	Sm.	Öl.	Gtl.	G. Sand.	Ög.	Vg.	Boh.	Dlsl.	Nrk.	Sdm.	Upl.	Vstm.	Vrm.	Dlr.	Gstr.	Hls.	Med.	Hrj.	Jmt.	Ång.	Vb.	Nb.	Ås. Lpm.	Ly. Lpm.	P. Lpm.	Lu. Lpm.	T. Lpm.
1																												
2																												
3	●	●	●	●		●			●			●	●	●		●		●				●						
4	●		●	●		●																						
5																												
6		●	●	●		●	●		●	●	●	●	●		●		●				●	●	●					
7																												
8																												
9																												
10																												
11																												
12																												
13																												
14																							●					
15																												
16																												
17		●	●	●		●	●	●			●	●	●	●	●	●	●	●	●	●	●	●	●					
18													●	●	●	●						●	●					
19	●	●	●	●	●															●								
20																												
21																												
22		●	●	●																								
23		●	●				●	●			●	●	●	●	●						●							
24																					●	●	●					●
25																												
26	●	●	●	●		●	●	●			●	●		●							●	●						
27																												
28																												
29	●																											
30																												
31																												
32			●	●	●	●																						
33																												
34																												
35																												

		Ø+AK	HE (s+n)	O (s+n)	B (ø+v)	VE	TE (y+i)	AA (y+i)	VA (y+i)	R (y+i)	HO (y+i)	SF (y+i)	MR (y+i)	ST (y+i)	NT (y+i)	Ns (y+i)
Scythris obscurella (Scop.)	1															
S. cuspidella (Den. & Schiff.)	2															
S. potentillella (Z.)	3	●			◗	●										
S. cicadella (Z.)	4															
S. bifissella (Hofm.)	5															
S. limbella (F.)	6	●	◖	◖		●	◖									
S. knochella (F.)	7															
S. scopolella (L.)	8															
S. paulella (HS.)	9															
S. clavella (Z.)	10															
S. seliniella (Z.)	11															
S. subseliniella (Hein.)	12															
S. sinensis (Feld. & Rog.)	13															
S. productella (Z.)	14															
S. palustris (Z.)	15															
S. muelleri (Mann)	16															
S. inspersella (Hb.)	17	●	◖													
S. noricella (Z.)	18			◖												
S. empetrella Karsh. & Niel.	19	●							◖							
S. siccella (Z.)	20															
S. tributella (Z.)	21															
S. picaepennis (Hw.)	22								◖							
S. disparella (Tgstr.)	23		◖													
S. fuscopterella Bengts.	24															
S. braschiella (Hofm.)	25															
S. laminella (Den. & Schiff.)	26	●	◖		◖	◖										
S. crassiuscula (HS.)	27															
S. ericivorella (Rag.)	28															
S. crypta Hann.	29															
S. restigerella (Z.)	30															
S. dissimilella (HS.)	31															
S. fuscoaenea (Hw.)	32															
S. grandipennis (Hw.)	33															
S. ericetella (Hein.)	34															
S. fallacella (Schl.)	35															

	Nn(ø+v)	TR(y+i)	F(v+i)	F(n+ø)	Al	Ab	N	Ka	St	Ta	Sa	Öa	Tb	Sb	Kb	Om	Ok	ObS	ObN	Ks	LkW	LkE	Le	Li	Vib	Kr	Lr
1											●				●												
2																											
3					●	●	●	●		●	●				●		●	●									
4																											
5																											
6					●	●	●	●	●	●	●	●						●									
7																											
8																											
9																											
10																											
11																											
12																											
13																											
14								●																			
15						●	●		●	●	●																
16																											
17					●	●	●		●	●	●	●			●			●	●								
18						●	●		●		●	●															
19						●	●										●	●									
20																											
21																											
22																											
23						●				●	●		●														
24																					●	●					
25																											
26						●	●				●			●	●		●										
27																											
28																											
29																											
30																											
31																											
32																											
33																											
34																											
35																											

References

Agenjo, R., 1971: Descrición de una nueva especie Madrileña del género *Scythris* Hb., 1816-1826, etc. - Eos, Madr., 46:7-14, pl.I.

Amsel, H.G., 1951: Una raccolta di Microlepidotteri della Dalmazia meridionale. - Redia, 36:411 -422.

Barnes, S.B. & McDunnough, J., 1917: Check list of Lepidoptera of boreal America. 392 pp. - Decatur.

Benander, P., 1951: *Scythris disparella* Tngstr. och *senescens* Stt. skilda arter (Lep.). - Opusc.ent., 16:191-192.

- 1965: Notes on larvae of Swedish Micro-Lepidoptera. II - Ibid., 30:1-23.

Bengtsson, B.Å., 1977: Two new species of Microlepidoptera from northern Sweden (Lepidoptera: Elachistidae, Scythrididae). - Ent.scand., 8:55-58.

Bradley, J.D., 1966: Type specimens of Microlepidoptera in the University Museum, Oxford, described by Haworth. - Entomologist's Gaz., 17:129-140.

Bruand, M.T., 1850: Catalogue systématique et synonymique des Lépidoptères du département du Doubs. Tinéides. - Mém.Soc.Émul.Doubs, (1)3(3):23-68.

Caradja, A., 1920: Beitrag zur Kenntnis der geographischen Verbreitung der Mikrolepidopteren des palaearktischen Faunengebietes nebst Beschreibung neuer Formen. - Dt.ent.Z.Iris, 34:75-179.

Christoph, H., 1882: Neue Lepidopteren des Amurgebietes (Forts.). - Bull. Soc.Nat. Moscou, 57: 5-47.

Common, I.F.B., 1970: Lepidoptera (Moths and Butterflies). Pp. 765-866 *in* Mackerras, I.M. (ed.): The Insects of Australia. xiii + 1029 pp., 8 pl. - Canberra.

- 1975: Evolution and Classification of the Lepidoptera. - A.Rev.Ent., 20:183-203.

Constant, M.A., 1865: Description de quelques Lépidoptères nouveaux. - Annls Soc.ent.Fr., ser.4, 5:189-198.

- 1885: Notes sur quelques Lépidoptères nouveaux. - Ibid., ser.6, 5:5-16.

Costa, O.G., (1829)-1886: Fauna dela Regno di Napoli. xii + 434 pp., 38 pls. - Napoli.

de Lesse, H. & Viette, P., 1949: Expeditions polaires françaises (Missions Paul-Emile Victor). Campagne 1949 au Groenland. Zoologie. - Première note: Microlepidoptera. - Annls Soc. ent.Fr., 115:81-92.

(Dennis, M & Schiffermüller, I.), 1775: Ankündung eines systematischen Werkes von den Schmetterlinge der Wienergegend. 323 pp., 3 pls.-Wien.

Duponchel, P.-A.-J., 1838-(1840). Pp. 332-333, pl.CCXCVIII *in* Godart, J.-B.: Histoire naturelle des Lépidoptères ou Papillons de France. 11, 720 pp., pls. 287-314. - Paris.

Döring, E., 1955: Zur Morphologie der Schmetterlingseier. 154 pp., 61 pls. - Berlin.

Emmet, A.M., 1979: A field guide to the smaller British Lepidoptera, 271 pp. - London.

Fabricius, J.C., 1775: Systema entomologiae. 832 pp. - Flensburgi et Lipsiae.

- 1794: Entomologia systematica emendata et aucta. 3, 349 pp. - Hafniae.

Falkovitsh, M.I., 1981: Scythrididae. Pp. 455-478 *in* Medvedeva, G.S.: Opredelitel nasekombych evropeijskoj tjasti SSSR, Tom IV, Tsheshoekrylye, Vtoraja tjast. 788 pp. - Leningrad. (Determination book on insects from the European part of SSSR, Part IV, Microlepidoptera, Second part. 788 pp. - Leningrad.).

Felder, R. & Rogenhofer, A.F., 1875: Lepidoptera *in:* Reise der österreichischen Fregatte Novara um die Erde in den Jahren 1857, 1858, 1859. Zoologische Teil. Zweiter Band, zweiter Abtheilung, Heft V, pl. 140, fig. 11. - Wien.

Filipjev, N., 1924: Microheterocera des Minussinsk Bezirks.I. - Ezheg.gosud. Muz.N.M.Mart' yanova, 2:42-43.

Forbes, W.T.M., 1923: The Lepidoptera of New York and neighboring States. - Mem.Cornell Univ.agric.Exp.Stn., 68:1-729.

Fuchs, A., 1901: Vier neue Kleinfalter der europäischen Fauna. - Stettin. ent.Ztg, 62:382-387.
- 1903: Neue Kleinfalter der europäischen Fauna - Jb.nassau.Ver.Naturk., 56:55-63.

Gerasimov, A., 1930: Zur Lepidopteren-Fauna Mittel-Asiens.I. - Annu.Mus. Zool.St.Peterbg., 31:21-48, 10 pls.

Glaser, W., 1962: *Scythris muelleri* Mn. - ein Neufund für das Burgenland. - Z.wien.ent.Ges., 47:137-138.

Hackman, W., 1945: Bidrag till kännedomen om våra *Scythris*-arter. - Notul. ent., 25:112-114.

Hannemann, H.J., 1958a: Beiträge zur Kenntnis einheimischer *Scythris*-Arten (Lep.Scythrididae). - Mitt.dt.ent.Ges., 17:65-67.
- 1958b: Beiträge zur Kenntnis einheimischer *Scythris*-Arten (Lep.Scythrididae). - Ibid., 17: 82-86.
- 1960: Zur Kenntnis der *Scythris*-Arten - Ibid., 19:84-87.
- 1961: Zoologische Ergebnisse der Mazedonienreisen Friedrich Kasys. II.Teil. Lepidoptera: Scythridae. - Sber.Akad.Wiss.Wien, Abt.1, 170:305-309.

Hartig, F., 1939: Su alcuni prototipi - Lepidotteri della collezione di Oronzio-Gabriele COSTA. - Annuar.R.Mus.zool.R.Univ.Napoli, 7:1-21.
- 1964: Microlepidotteri della Venezia Tridentina e delle regioni adiacenti. Parte III. (Fam.Gelechiidae-Micropterygidae). - Studi trent.Sci.nat., 41:1-292.

Haworth, A.H., 1803-1828: Lepidoptera Britannica. 1-4, xxxvi + 610 pp. - Londini.

Heinemann, H.v., 1872: Eene nieuwe soort van *Butalis*. - Tijdschr.Ent., 15:280-284.
- & Wocke, M.F., 1877: Die Schmetterlinge Deutschlands und der Schweiz. 2.Abt., Kleinschmetterlinge. 2, vi + 825 + 102 pp. - Braunschweig.

Hemming, F., 1937: Hübner. A bibliographical and systematic account on the entomological works of Jacob Hübner and the supplements by Carl Geyer, G.F.v.Fröhlich and G.A.W.Herrich-Schäffer. Vol.I, xxxiv + 605 pp., vol.II, ix + 274 pp. - London.

Heppner, J.B., 1977: The status of the Glyphipterigidae and a reassessment of relationships in Yponomeutoid families and Ditrysian superfamilies. - J.Lepid.Soc., 31:124-134.

Hering, M., 1918: Zur Biologie und systematischen Stellung von *Scythris temperatella* Led. - Dt.ent.Z.Iris, 32:122-129.
- 1924: Beitrag zur Kenntnis der Microlepidopteren-Fauna Finlands. - Notul. ent., 4:75-84.
- 1932: Die Schmetterlinge nach ihre Arten dargestellt. *In* Brohmer, P., Ehrmann, P. & Ulmer, G.: Die Tierwelt Mitteleuropas. Ergänzungsband 1, 545 pp. - Leipzig.

Herrich-Schäffer, G.A.W., 1847-1855: Systematische Bearbeitung der Schmetterlinge von Europa.5, 394 pp., pls. 1-124 (Tineidae), 1-7 (Pterophorides), 1 (Micropteryges). - Regensburg.

Hofmann, O., 1888: Beiträge zur Kenntnis der Butaliden. - Stettin.ent.Ztg, 49:335-347, 1 pl.
- 1889: *Butalis bifissella* n.sp. und *Lypusa? fulvipennella* m. - Ibid., 50:107-110.
- 1893: Beiträge zur Naturgeschichte der Tineinen. - Ibid., 54:308-309.
- 1897: Eine neue *Butalis*-Art. - Dt.ent.Z.Iris, 10:241-244.

Hübner, J., 1796-(1836): Sammlung europäischer Schmetterlinge. 8, 78 pp. (1796), 71 pls. - Augsburg.
- 1816-(1825): Verzeichniss bekannter Schmettlinge (sic). 431 pp. - Augsburg.

International Code of Zoological Nomenclature adopted by the XV International Congress of Zoology, 1964. - London.

Joannis, J.de, 1909: Une nouvelle espèce de *Scythris* (Microlép.) des environs de Vannes. - Bull. Soc.ent.Fr., 1908:248-250.

- 1920a: Note sur une petite collection de Microlépidoptères provenent de St. Saens (Seine-Inférieure) et description d'espèces nouvelles. - Ibid., 1920:142-147.

- 1920b: Remarques sur la définition de genre *Scythris* Hb. (Lep). - Ibid., 1920:170-173.

Johansson, R., 1971: Notes on Nepticulidae (Lepidoptera) I. A revision of the *Nepticula ruficapitella* Group. - Ent.scand., 2:241-262.

Jäckh, E., 1977: Bearbeitung der Gattung *Scythris* Hübner (Lepidoptera, Scythrididae). 1. Die 'grandipennis-Gruppe'. - Dt.ent.Z.(N.F.), 24:261-271, 11 pls.

- 1978a: Bearbeitung der Gattung *Scythris* Hübner (Lepidoptera, Scythrididae). 2. Eine neue *Scythris*-Art aus Spanien: *Scythris limbelloides* n.sp. - Z.ArbGem.öst.Ent., 29:81-84.

- 1978b: Bearbeitung der Gattung *Scythris* Hübner (Lepidoptera, Scythrididae). 3. Arten mit einer weissen Längsstrieme. - Dt.ent.Z.(N.F.), 25:71-89, 21 pls.

- 1978c: Bearbeitung der Gattung *Scythris* Hübner (Lepidoptera, Scythrididae). 4. Unbeschriebene Arten aus Italien. - Boll.Mus.civ. Storia nat. Verona, 5:1-14, 5 pls.

Karlsholt, O. & Nielsen, E. Schmidt, 1976a: Systematisk fortegnelse over Danmarks sommerfugle (A Catalogue of the Lepidoptera of Denmark). 128 pp. - Klampenborg.

- 1976b: *Nemapogon wolffiella* nom.nov. for *N.albipunctella* (Haworth) and *Scythris empetrella* nom.nov. for *S.variella* (Stephens) (Lepidoptera: Tineidae and Scythrididae). - Ent. scand., 7:151-152.

Kasy, F., 1962: Zwei neue *Scythris*-Arten aus Südwesteuropa. - Annln naturh. Mus.Wien, 65: 167-171.

Klimesch, J., 1951: Über eine neue *Scythris*-Art aus den Nordost-Alpen *(Scythris saxicola* sp.n.: Lep., Scythrididae). - Z.wien.ent.Ges., 36:141-144, 1 pl.

Krone, W., 1905: P.100 *in* Jahres-Bericht des Wiener entomologisches Vereines, 1904. - Wien.

Kyrki, J., 1978: Suomen pikkuperhosten levinneisyys.1. Luonnontieteellisten maakuntien lajisto (Lepidoptera: Micropterygidae-Pterophoridae). - Notul. ent., 58:37-67.

Leraut, P., 1980: Liste systématique et synonymique des Lépidoptères de France, Belgique et Corse. 334 pp. - Paris.

Lhomme, L., 1935-(1963): Catalogue des Lépidoptères de France et de Belgique. 2. 1253 pp. - Douelle (Lot.).

Lid, J., 1974: Norsk og Svensk Flora. 808 pp. - Oslo

MacKay, M.R., 1972: Larval sketches of some Microlepidoptera, chiefly North American. - Mem.ent.Soc.Can., 88:1-83.

MacKechnie-Jarvis, C., 1975: (As President in the chair exhibiting different observations). - Proc. Brit.ent.nat.Hist.Soc., 7:91.

Mann, J., 1871: Beschreibung drei neuer Kleinschmetterlinge. - Verh.zool.-bot.Ges.Wien, 21:80 -82.

Marchand, S. le, 1938: *Scythris joannisella*, nova species. - Revue fr. Lépidopt., 9:131-136.

Matsumura, S., 1931: 6000 Illustrated Insects of Japan-Empire, pp. 1094-1095. - Tokyo.

McDunnough, J., 1939: Check List of the Lepidoptera of Canada and the United States of America. Part II. Microlepidoptera. - Mem.sth.Calif.Acad.Sci.,2(1), 171 pp.

Meyrick, E., 1895: Handbook of British Lepidoptera. 843 pp. - London.

- 1928: A revised handbook of British Lepidoptera. vi+914 pp. - London.

- 1929a: A new British species of Micro-Lepidoptera. - Entomologist, 62:149-150.

- 1929b: Exotic Microlepidoptera, 3. - Marlborough.

132

Millière, P., 1876: Cataloque raisonné des Lépidoptères du département des Alpes-Maritimes. Part 3: (249)-455, pls.I,II. - Paris.

Mosher, E., 1916: A classification of the Lepidoptera based on characters of the pupa. - Bull.Ill. St.Lab.nat.Hist., 12:15-159.

Opheim, M., 1978: The Lepidoptera of Norway. Check-List. Part III Gelechioidea (first part). - Trondheim.

Osthelder, L., 1951: Die Schmetterlinge Südbayerns und der angrenzenden nördlichen Kalkalpen. II.Teil. Die Kleinschmetterlinge. - Beilage Mitt.münch.ent.Ges., 41:113-250.

Pallesen, G. & Palm, E., 1974: Fund af småsommerfugle fra Danmark i 1973. - Flora og Fauna, 80:95-101.

- 1975: Fund af småsommerfugle fra Danmark i 1974. - Ibid., 81:73-78.

Park, K.-T., 1977: Discovery of Scythris sinensis (F.&R.) from Korea (Lepidoptera, Scythrididae). - Korean J.Entomol., 7:1-2.

Passerin d'Entrèves, P., 1977: Revisione degli Scitrididi (Lepidoptera, Scythrididae) paleartici. III. - Le specie de Scythris descritte da H.G.Amsel, J.Klimesch, J.Müller-Rutz e A.Rössler. - Boll.Mus.Zool.Univ.Torino, 1977, no.5:57-76.

- 1979: Revisione degli Scitrididi (Lepidoptera, Scythrididae) paleartici. IV. - I tipi di Scythris dell'Instituto Español de Entomología di Madrid. - Ibid, 1979, no.3:83-90.

- 1980: Revisione degli Scitrididi (Lepidoptera, Scythrididae) paleartici. V. - I tipi de Scythris del Naturhistorisches Museum di Vienna. - Ibid., 1980, no.5:41-60.

Petersen, W., 1923: Lepidopteren-Fauna von Estland, I & II, 588 pp. - Reval.

Pierce, F.N. & Metcalfe, J.W., 1935a: The genus Scythris, introducing S.fallacella Schläg., a species new to Britain. - Entomologist, 68:45-51.

- 1935b: The genitalia of the Tineid families of the Lepidoptera of the British Islands. xxii + 116 pp., 68 pls. - E.W.Classey Ltd, Middlesex (Reprint).

Powell, J.A., 1975: A remarkable new genus of brachypterous moth from coastal sand dunes in California (Lepidoptera: Gelechioidea, Scythrididae). - Ann.ent.Soc.Am., 69:325-339.

Ragonot, E.L., 1881: (Descriptions de trois nouvelles espèces de tinéites du genre Butalis, etc.). - Annls Soc.ent.Fr., Ser.5, 10 (1880):CXX-CXXII.

Rapp, O., 1936: Beiträge zur Fauna Thüringens, 2. Microlepidoptera, Kleinschmetterlinge (1), 240 pp. - Erfurt.

Rebel, H., 1900: Neue palaearctische Tineen. - Dt.ent.Z.Iris, 13.

- 1901. In Staudinger, O. & Rebel, H.: Catalog der Lepidopteren des palaearktischen Faunengebietes. II. Theil. Famil. Pyralidae-Micropterygidae. 368 pp. - Berlin.

- 1918: Lepidopteren aus Mittelalbanien. - Z.öst.EntVer., 3:75-77 & 85-88.

- 1938: Zwei neue Arten von Scythris Hbn. aus Polen (Elachistidae, Lep.). - Annls Mus.zool. pol., 13, (no.10):105-108, pl.VIII.

Robinson, G.S. & Nielsen, E.S., 1983: The Microlepidoptera described by Linnaeus and Clerck. - Syst.Ent., 8:191-242.

Romaniszyn, J., 1929-1930: in Schille: Fauna motyli Polski. Prace Monogr. Kom.Fizj. PAU, Kraków, 7, 558 pp.

Rössler, U., 1864-66: Verzeichniss der Schmetterlinge des Herzogthums Nassau. - Jb.nass.Ver. Naturk., 19.&20.Hft., pp.99-442.

Samuelsson, G., 1976: Ett säkert fynd av Scythris knochella F. i Sverige (Lep.Scythridae). - Entomologen, 5:17-18.

Sattler, K., 1971: On Scythris sinensis (Felder & Rogenhofer) and S.chrysopygella Caradja (Lepidoptera, Scythrididae). - Reichenbachia, 14:39-45.

- 1981: Scythris inspersella (Huebner, (1817)) new to the British Fauna (Lepidoptera:Scythridi-

133

dae). - Entomologist's Gaz., 32:13-17.

Sattler, K. & Tremewan, W.G., 1978: A supplementary catalogue of the family-group and genus-group names of the *Coleophoridae* (Lepidoptera). - Bull.Br.Mus.nat.Hist.Ent. 37:73-96.

Schläger, F., 1847: Entdeckungen, Berichtungen und sonstige Bemerkungen. - Ber.lepidopt. Tausch-Ver., 1847:223-244.

Schmidt, A.v., 1941: Neue spanische Microlepidopteren. - Boln R.Soc.esp.Hist.nat., 38 (1940): 37-39, pl.II.

Schütze, K.T., 1897: Mittheilungen über einige Kleinschmetterlinge. - Stettin. ent.Ztg, 58:299-314.

- 1904: Mitteilungen über einige Kleinschmetterlinge. - Dt.ent.Z.Iris, 17:192-208.

- 1931: Die Biologie der Kleinschmetterlinge unter besonderer Berücksichtigung ihrer Nährpflanzen und Erscheinungszeiten. 235 pp. - Frankfurt am Main.

Scopoli, J.A., 1763: Entomologia carniolica exhibens insecta carnioliae indigene et distributa in ordines, genera, species, varietates methodo Linnaeana. 421 pp. - Vindobonae.

Sheppard, A.C., 1974: Palaearctic Lepidoptera in the Province of Québec. - Ann.ent.Soc.Queb., 19:119-120.

Spuler, A., 1892: Zur Phylogenie und Ontogenie des Flügelgeäders der Schmetterlinge. - Z.wiss. Zool., 53:597-646, pls.XXV-XXVI.

- 1903-1910: Die Schmetterlinge Europas. 2, 523 pp. - Stuttgart.

Stainton, H.T., 1850: On *Elachista aeratella*, Zeller, and several species with which it is likely to be confounded. - Trans. ent.Soc.Lond., 1:21-25.

- 1854: Insecta Britannica. Lepidoptera: Tineina. viii + 313 pp., 10 pls. - London.

Staudinger, O., 1859: Diagnosen nebst kurzen Beschreibungen neuer Andalusischer Schmetterlinge. - Stettin.ent.Ztg, 20:211-259.

- 1870: Beitrag zur Lepidopterenfauna Griechenlands. - Horae Soc.ent.ross., 7:3-304.

Stephens, J.F., 1834-1835: Illustrations of British Entomology. Haustellata 4, 436 pp., pls. 33-40. - London.

Steudel, W. & Hofmann, E., 1882: Verzeichniss württembergischer Kleinschmetterlinge. - Jh.Ver. vaterl.Naturk.Württ., 38:143-262.

Stitz, H., 1900: Der Genitalapparat der Mikrolepidopteren. 1, der männliche Genitalapparat. - Zool.Jb. (Anat.Ontog.), 14:135-176, pls. VII-XI.

Šulcs, A., 1973: Neue und wenig bekannte Arten der Lepidopteren-Fauna Lettlands. 5.Mitteilung. - Annls Ent.Fenn., 39:1-16.

- 1981: Neue und wenig bekannte Arten der Lepidopteren-Fauna Lettlands. 8.Mitteilung. - Notul.ent., 61:91-102.

Svensson, I., 1978: Anmärkningsvärda fynd av Microlepidoptera i Sverige 1977. - Ent.Tidskr., 99:87-94.

- 1979: Anmärkningsvärda fynd av Microlepidoptera i Sverige 1978. - Ibid., 100:91-97.

Tengström, J.M.J., 1848: Bidrag till Finlands Fjäril-Fauna. - Notis. Sällsk. Faun.Fl.fenn.Förh., 1:69-164.

Thunberg, C., 1794: D.D. Dissertatio entomologica sistens Insecta svecica 7. - Upsaliae.

Traugott-Olsen, E. & Nielsen, E. Schmidt, 1977: The Elachistidae (Lepidoptera) of Fennoscandia and Denmark. - Fauna ent.scand., 6, 299 pp., 108 pls.

Treitschke, F., 1833: Die Schmetterlinge von Europa. 9(2). - Leipzig.

Tremewan, W.G., 1977: The publications on Lepidoptera by O.G. and A. Costa and the nominal taxa described therein. - Bull.Br.Mus.nat.Hist.(hist.Ser.), 5(3):213-232.

Turati, E., 1924: Spedizione Lepidotterologica in Cirenaica, 1921-1922. - Atti Soc.ital.Sci.nat., 63:21-191.

Turner, A.J., 1947: A Review of the phylogeny and classification of the Lepidoptera. - Proc.Linn. Soc.N.S.W., 71:303-338.

Wallengren, H.D.J., 1875: Species Tortricum et Tinearum Scandinaviae. - Bih.K.svenske Vetensk Akad.Handl. 3(5):1-90.

Wocke, M.F., 1877: see Heinemann, 1877.

Wollf, N.L., 1959: Notes on *Scythris cicadella* Zell. (Lep.Scythrididae). - Opusc.ent., 24:1-2.

- 1964: The Lepidoptera of Greenland. - Meddr Grønland, 159 (11):74 pp., 21 pls.

Zander, E., 1905: Der männliche Genitalapparat der Butaliden. - Z.wiss.Zool., 79:308-323.

Zeller, P.C., 1839: Versuch einer naturgemässen Eintheilung der Schaben. - Oken, Isis, 1839: 167-220.

- 1843: Nachricht über eine lepidopterologische Excursion von Wien aus in die steyrischen Alpen. - Stettin.ent.Ztg., 4:144-151.

- 1847: Bemerkungen über die auf einer Reise nach Italien und Sicilien gesammelten Schmetterlingsarten. - Oken, Isis, 1847: 829-835 (part).

- 1855: Die Arten der Gattung *Butalis* beschrieben. - Linnaea Ent., 10: 196-267, 2 pls.

Zimmerman, E.C., 1978: Microlepidoptera II. Gelechioidea. - Insects of Hawaii, 9:883-1903.

Index

Page references are given to the keys, the main treatment (in bold) and the three sets of illustrations. For these 'i' indicates colour drawing of the imago, 'm' is male genitalia, and 'f' is female genitalia. Synonyms are given in italics.

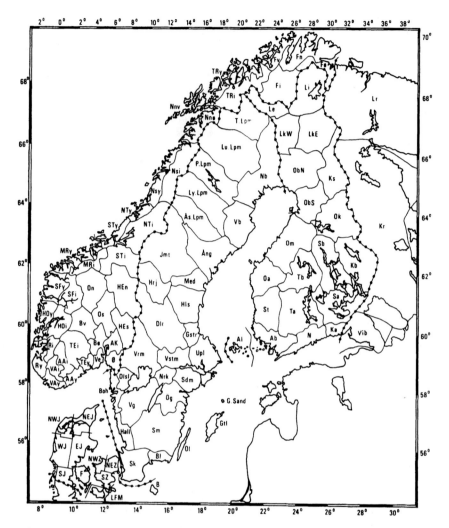

List of abbreviations for the provinces used throughout the text,
on the map and in the following tables.

DENMARK

SJ	South Jutland	LFM	Lolland, Falster, Møn
EJ	East Jutland	SZ	South Zealand
WJ	West Jutland	NWZ	North West Zealand
NWJ	North West Jutland	NEZ	North East Zealand
NEJ	North East Jutland	B	Bornholm
F	Funen		